UK Wind Energy Techn

T0231628

Phase 1 of the EPSRC SUPERGEN Wind programme began in March 2006 and work continued under Phase 2 until March 2014. The strategic aim was to re-establish a strong research community in wind energy technologies, across the UK's leading academic and industrial research organisations.

UK Wind Energy Technologies gives a comprehensive overview of the range of wind energy research undertaken in the UK under Phases 1 and 2 to achieve this goal. Specific topics covered in the book include: wind resource assessment, turbine array layout, environmental interactions, control of turbines, drive train reliability and condition monitoring, turbine array electrical connection, power transmission to grid, assessment of operations and maintenance strategies, and the analysis of turbine foundations and structures. Since the completion of Phase 2 the SUPERGEN Wind Consortium partners have formed a networking Hub, which is now the principal national coordinating body for academic research into wind energy in the UK.

This book will be of interest to researchers and engineers from industry and academia and also provides workers from other countries with an overview of the range of activity within the UK resulting from the SUPERGEN Wind programme to date.

Simon Hogg is Professor in the School of Engineering and Computing Sciences and DONG Energy Professor in Renewable Energy at Durham University, UK. He was the PI of the £4.8m EPSRC SUPERGEN Wind 2 Consortium and holds a number of industrially-funded research grants on new turbine technologies.

Christopher J. Crabtree is a Lecturer in Wind Energy Systems in the School of Engineering and Computing Sciences, Durham University, UK. He acted as project management assistant to the EPSRC SUPERGEN Wind Energy Technologies Consortium.

UK Wind Energy Technologies

Edited by Simon Hogg and
Christopher J. Crabtree

LONDON AND NEW YORK

First published 2017
by Routledge
4 Park Square, Milton Park, Abingdon, Oxon OX14 4RN

605 Third Avenue, New York, NY 10017

Routledge is an imprint of the Taylor & Francis Group, an informa business

First issued in paperback 2018

British Library Cataloguing in Publication Data
A catalogue record for this book is available from the British Library

Library of Congress Cataloging in Publication Data
Names: Hogg, Simon (Simon I.), editor. | Crabtree, Christopher (Christopher James), editor.
Title: UK wind energy technologies / edited by Simon Hogg and Christopher Crabtree.
Other titles: United Kingdom wind energy technologies
Description: Abingdon, Oxon ; New York, NY : Routledge, 2016. | Includes bibliographical references.
Identifiers: LCCN 2016005211 | ISBN 9781138780460 (hb) | ISBN 9781315681382 (ebook)
Subjects: LCSH: Wind power--Great Britain.
Classification: LCC TJ820 .H335 2016 | DDC 621.31/21360941--dc23
LC record available at http://lccn.loc.gov/2016005211

ISBN: 978-1-138-78046-0 (hbk)
ISBN: 978-1-138-39342-4 (pbk)

Typeset in Goudy
by HWA Text and Data Management, London

Dedicated to

Prof. Peter Tavner

who was instrumental, along with Prof. Bill Leithead and other consortium members, in re-establishing significant wind energy research activity in UK universities through the SUPERGEN Wind Consortium.

Contents

Figures

Tables

Contributors

Olimpo Anaya-Lara is a reader in electronic and electrical engineering at the University of Strathclyde. His research is in power system dynamics and the modelling and control of wind power plant. He is part of the management team of the EPSRC Centre for Doctoral Training in Wind and Marine Energy Systems and leads Strathclyde's contribution to the Wind Integration Sub-Programme of the EERA Joint Programme Wind. During 2010-2011, he was visiting professor at NTNU, Trondheim, funded by Det Norske Veritas, and has published numerous papers and books on wind energy electrical systems.

Antony Beddard graduated from the University of Manchester in 2009 with a MEng electrical and electronic engineering degree. Following work experience with AREVA T&D, he has been undertaking a PhD at the University of Manchester. He is currently in his third year of his PhD, investigating the use of VSC-HVDC for offshore wind farm connection as part of the SUPERGEN Wind consortium. He has a keen interest in the area of HVDC and the importance of this technology for present and future applications.

Mike Barnes is a professor in the Power Conversion group in the School of Electrical and Electronic Engineering at the University of Manchester . He received BEng and PhD degrees from the University of Warwick, UK in 1993 and 1998 respectively, where he was also a research associate and then lecturer. He joined the University of Manchester Institute of Science and Technology (UMIST, now the University of Manchester) in 1997 as a lecturer in power electronics, and has developed research interests in high voltage DC transmission, offshore wind energy and flexible AC transmission systems. He is an associate editor of the *IEEE Transactions on Energy Conversion*.

Anthony K. Brown is Professor of Communication Engineering at the University of Manchester. Tony is a Fellow of the IET and the IMA, a senior member of the IEEE and a member of the IOD. He joined academia in 2003 as research group leader after a 30-year industrial career including senior board level positions. He maintains strong industrial links and is currently CTO and director of two companies. Tony's research interests are principally in radio astronomy instrumentation, radar and related imaging, and communications in complex environments

Ian Cotton first studied at the University of Sheffield. After working for an electricity distribution company he returned to UMIST (now the University of Manchester) to complete a PhD on lightning protection of wind turbines. He is currently Professor of High Voltage Technology in the School Of Electrical and Electronic Engineering at the University of Manchester. With research interests in novel power system equipment, power systems transients and power system induced corrosion. He is Director of Manchester Energy, a member of the IET, a senior member of the IEEE and a chartered engineer.

Christopher J. Crabtree is a Lecturer in Wind Energy Systems at Durham University, UK. He received his MEng degree in engineering from Durham in 2007 and his PhD degree from Durham in 2011 with a thesis on condition monitoring techniques for wind turbines. His research interests include reliability, condition monitoring, operations and maintenance of offshore wind farms. Much of his research has been undertaken as part of the UK EPSRC SUPERGEN Wind Energy Technologies Consortium.

Laith Danoon is a lecturer at the School of Electrical and Electronic Engineering, the University of Manchester. Dr Danoon is an award winning engineering and research scientist specialising in radar and propagation modelling. His work on the propagation and multiple reflections of radar waves within complex environments such as wind farms has been internationally recognised sparking interest within the research communities as well as various industrial partners within the wind industry.

Geoff Dutton led the wind energy research group of the Energy Research Unit (ERU) at STFC Rutherford Appleton Laboratory until 2014. His research interests include non-destructive testing techniques for wind turbine blades, fatigue of blade materials, blade (finite element) modelling, and dynamic fluid-structure blade loading. He has worked within large research projects including the EU-funded OPTIMAT Blades and the UPWIND project He is also interested in the efficient management of energy data and was responsible for setting up and managing the UKERC Energy Data Centre. He is a chartered engineer and member of the Institution of Mechanical Engineers.

Philip Hancock is a Reader in Fluid Mechanics in the Department of Mechanical Engineering Sciences at the University of Surrey. His particular research interests are in aerodynamics, turbulence and boundary layer meteorology; wind-tunnel simulation of neutral, stable and convective atmospheric boundary layer flows; wind turbine wakes, wake-wake and wake turbine interactions; structure of complex turbulent flows, separated flows and wakes; aerodynamic and bluff body flows, and wind power aerodynamics.

Simon Hogg is Professor in the School of Engineering and Computing Sciences and DONG Energy Professor in Renewable Energy at Durham University, UK and is Executive Director of the Durham Energy Institute. His research interests include conventional steam and gas turbine power plant, wind turbines, energy systems and waste heat recovery. He is principal investigator of the EPRSC Future Conventional Power Consortium and sits on the EPSRC SUPERGEN Wind Hub Executive Committee. Simon graduated with a BSc and PhD from the University of Manchester, funded by Rolls Royce to develop computational fluid dynamics codes for gas turbine combustor design. He held university posts at Oxford and Leicester before joining Alstom's power generation division in 1998, becoming engineering director in 2007. Simon is a chartered engineer and fellow of the Institution of Mechanical Engineers.

Behzad Kazemtabrizi obtained his PhD in power systems from Glasgow University in 2011. He is a lecturer in electrical engineering at Durham University, after joining Durham as a research associate in 2012. His research interests include advanced electrical power systems modelling, simulation, analysis and optimisation for improved performance and reliability with a focus on renewable energy resources, in particular wind. His research extends into wind farm asset management and integration through system reliability and energy production simulation.

William E. Leithead is Professor of Systems and Control at the University of Strathclyde. He has been chair of the SUPERGEN Wind consortium since its inception in 2006. He is involved with many national and international wind energy research and policy committees including the European Academy of Wind Energy Executive, EERA Joint Programme Wind Steering Committee and ORECatapult Research Advisory Group. His research is in control engineering and wind energy systems, with particular focus on the conceptual design of wind turbines and wind turbine and farm control systems. He is Director of the EPSRC Centre for Doctoral Training in Wind Energy Systems and its successor, the EPSRC Centre for Doctoral Training in Wind and Marine Energy Systems

Antonio Luque received his first degree from Terrassa School of Engineering (Spain) in 2007 before joining the University of Strathclyde to complete his MSc and PhD degrees within electrical and electronics engineering. He is currently a research assistant at the University of Reading where his research interests include power network modelling for renewable energy integration, control systems for energy storage and network data analysis.

Vidyadhar Peesapati is a research fellow in the Electrical Energy and Power Systems Group at the University of Manchester. He joined the University of Manchester in 2005, to pursue a Master's degree in electrical power systems and then a PhD in high voltage electrical energy. He is presently working with different companies and research groups in developing innovative, cost effective and sustainable solutions for energy transmission, generation and storage.

Alan Ruddell is a research engineer at STFC Rutherford Appleton Laboratory and has worked for over 20 years in renewable energy systems research, including wind energy systems and energy storage. He is the author or co-author of over 40 publications, including ten journal papers and two book chapters on energy storage, and the UK Energy Research Centre (UKERC) Wind Energy and Energy Storage landscapes.

Adam Stock is currently with SgurrControl having been a research associate in the Industrial Control Centre at the University of Strathclyde, where he completed a PhD in augmented control for wind turbines. His research interests include wind turbine control, wind farm control, and flexible operation of wind power plant. Prior to working at the University of Strathclyde (from the end of 2010), he obtained his MEng in Mechanical Engineering from the University of Newcastle Upon Tyne (in 2007), before working in Industry.

Peter J. Tavner is an Emeritus Professor of the School of Engineering and Computing Sciences, Durham University. He received degrees from Cambridge University and Southampton University, and latterly the DSc from Durham University (2012). Following his PhD he worked for the UK electricity supply industry in senior research, development and technical positions, most recently as Group Technical Director of FKI Energy Technology. He joined Durham University in 2003 and was principal investigator of the SUPERGEN Wind Consortium and the Sino-British Future Renewable Energy Network Systems (FRENS) Consortium, and a member of the SUPERGEN Marine and EU FP7 ReliaWind Consortia. He is a fellow of the Institution of Engineering Technology and winner of the Institution Premium of the IET.

Simon Watson is Professor of Wind Energy in the Centre for Renewable Energy Systems Technology at Loughborough University. He has been working at Loughborough University since 2001 and his main research areas include condition monitoring of wind turbines, wind resource assessment, wind power forecasting, wake modelling, wave power device modelling and climate change impacts. He is a member of the European Academy for Wind Energy and is on the editorial panel for *Wind Engineering*.

Richard Williams graduated in 2004, with an MEng in mechanical engineering from Durham University. Before and during his degree he undertook an apprenticeship with Black & Decker. After a year with Alstom Power, Richard returned to Durham to complete his PhD entitled 'Large tip clearance flows in high pressure stages of axial compressors'. He is a post-doctoral researcher at Durham University using numerous experimental, computational and analytical techniques. His research interests include steam turbine aerodynamics, steam turbine secondary flow path sealing, combined heat and power systems, skin friction measurements and asset management of wind turbines.

Donatella Zappalá is a research associate at Durham University. She earned the MSc degree at Universitá di Perugia (Italy) in 2008 before joining the School of Engineering and Computing Sciences at Durham where she received the MSc degree in new and renewable energy in 2011 and the PhD degree in advanced condition monitoring techniques for wind turbines in 2015. Her current research includes wind turbine condition monitoring and reliability and focuses on the development of novel techniques for drive train electrical and mechanical fault detection, working both experimentally and with industrial partners.

Preface

This book provides an overview of the status of wind energy research in the UK, following completion of the first two phases of the UK Government funded EPSRC SUPERGEN Wind programme. Phase 1 started in March 2006 and the work continued under Phase 2 until September 2014. The programme was a direct result of UK Government (EPSRC) recognising that there had been little research funding in the UK for early Technology Readiness Level developments in wind for over a decade prior to this. This SUPERGEN programme was therefore launched with two principal strategic objectives in mind. These were to re-establish a wind technology academic and industry research network in the UK and to contribute to the over-riding technical objective of reducing the cost of energy from wind.

The book covers the full range of engineering topics researched under Phases 1 and 2 of the programme. These range from research topics that are directly related to harvesting energy from the wind using turbines, through to the development of supporting technologies such as transmission and connection to the National Grid. Specifically, the topics included in the book are: wind resource assessment, turbine array layout, environmental interactions, control of turbines, drive train reliability and condition monitoring, turbine array electrical connection, power transmission to grid and assessment of operations and maintenance strategies. The work summarised in the book and the extensive base of references to technical outputs from the SUPERGEN programme demonstrate that the principal objectives of the programme have been successfully realised during Phases 1 and 2.

During 2014, the SUPERGEN Wind programme transitioned to Hub status. The EPSRC has funded a Hub for five years to 2019 in the first instance. The aim of the Hub is to continue to develop the important academic, industrial and policy linkages that were established during the earlier phases of the programme and to own the technology strategy for driving research in wind forward and for exploiting the research outcomes.

This book will be of interest to energy researchers and engineers from industry, academia and government who have interests in evaluating prospective wind farm sites, developing onshore and offshore wind farms, operating and maintaining wind farms, improving the reliability of turbines and extending their operational lives. It is accessible to professionals across all sectors of the wind industry from early career researchers through to senior industrial project managers. Whilst the book is not specifically designed as a course textbook, it is structured in a form such that individual chapters could be adopted as required to support and provide supplementary reading for specialist courses. These might include courses on energy generation/conversion, electrical machines, materials and control of turbines, to name but a few. The book will also provide workers from other countries with an overview of the range of activity within the UK resulting from Phases 1 and 2 of the SUPERGEN Wind programme.

Simon Hogg
Christopher J. Crabtree

1 Wind resource

Simon Watson and Philip Hancock

This chapter considers the offshore wind resource and how it is likely to be translated into power production by large arrays of offshore wind turbines. Firstly, the characteristics of the offshore wind resource were studied using long-term reanalysis data from the ERA–40 dataset. For two offshore sites, a more in-depth prediction of the wind resource was made using a mesoscale model. Finally, the results of two studies of the characteristics of offshore wind farm wakes is presented, using a computational fluid dynamics (CFD) model and then inferred from scale model measurements in a meteorological wind tunnel.

Characteristics of the UK offshore wind resource

In recent years, there has been an increased interest in assessing the UK offshore wind resource as offshore wind farms have started to be planned and developed. The optimal way of assessing a site is by erecting a mast and making measurements at several heights over a period of greater than a year. The drawback of this approach is the high cost of masts capable of making measurements over the extent of a wind turbine rotor. In addition, such masts can only make measurements at one geographical location. In order to help address the second point, a study has been made of the spatial variation in long-term wind resource for the UK Round 3 offshore wind farm sites using a popular reanalysis dataset, namely ERA–40. The characteristics of the wind resource are assessed to give an insight into the wind conditions likely to be experienced at future Round 3 wind farm sites.

The ERA–40 database: background

Reanalysis data are a source of meteorological information derived from numerical weather prediction (NWP) models initialised using historical weather observations from satellites, aircraft, balloons, and surface stations. The NWP model generates time series, which can cover several decades, of gridded atmospheric variables containing among others: temperature, pressure, wind, humidity and precipitation.

In particular, ERA–40 is a second generation reanalysis of the global atmospheric data over a 45-year period from 1st September 1957 to 31st August

2002 (Uppala 2005). It was created from an ECMWF Integrated Forecast Model and it was the first reanalysis to directly assimilate satellite radiance data as well as the Cloud Motion Winds from 1979 onwards. The model has 60 vertical levels with a spectral resolution of around 125 km (T159) and observations recorded every 6-hours through a 3D variational analysis.

This section covers research undertaken using 21 years of this dataset, corresponding to the data from 1st January 1968 to 31st December 1989. The wind data were available in the form a geographical grid with a horizontal resolution of 1° × 1° every 6 hours at 10 metres height.

In this section, the wind conditions are summarised at locations representative of the original nine UK Round 3 Offshore Wind Farm development sites. The summary statistics presented are designed to give an idea of the variation in wind speed and wind speed distribution at the sites as a function of direction, season, and time of day. The data analysed are 10 m values so would need to be extrapolated to wind turbine hub height for energy yield analysis.

The UK Round 3 offshore regions

This section gives relevant information concerning the UK Round 3 sites. Representative locations in each of the Round 3 sites were identified and data interpolated using a bilinear interpolation based on the four closest grid points from the ERA–40 reanalysis dataset. A representative centre point for each Round 3 site was chosen using maps provided by the Crown Estate. Figure 1.1 shows these maps and the particular details for each site. The specific coordinates of the nine representative locations are listed in Table 1.1.

Wind direction

This section looks at the behaviour of the frequency distribution of the wind for the Round 3 sites. First, the results are presented using wind roses and then in order to look at temporal variation in the wind direction, data are presented as frequency plots.

The wind roses are presented in Figure 1.2 showing the directional pattern for each interpolated Round 3 site over the whole period and by season (Winter = DJF, Spring = MAM, Summer = JJA, Autumn = SON). Each direction of the wind rose has superimposed the distribution of wind speed in bins of 5 m/s.

It can be seen from examination of these wind roses that the sites show a broadly similar behaviour with a predominance of winds from the south through to west sectors. However, those sites relatively close to the south coast, i.e. West Isle of Wight and Southern Array, show slightly different behaviour with evidence of channelling along the south-west to north-east axis. This would seem to be consistent with the location of the land masses of the south-coast of England, the Isle of Wight and the northern French coast with the effect of the English Channel noticeable.

The wind roses also show some differences by season. Aside from the obvious differences in mean wind speed (highest in winter and lowest in summer), some

Table 1.1 Coordinates (in degrees) for the interpolated centre of each Round 3 site (tIP) and the four surrounding coordinates from the ERA–40 reanalysis database (t1, t2, t3, t4)

Round 3 sites		t1		t2		t3		t4		tIP	
		Lat	Lon	Lat	Lon	Lat	Lon	Lat	Lon	Lat	Lon
1	Moray Firth	58	–2	58	–3	59	–3	59	–2	58.156	–2.94
2	Firth of Forth	56	–1	56	–2	57	–2	57	–1	56.313	–1.75
3	Dogger Bank	55	3	55	2	56	2	56	3	55.375	2.625
4	Hornsea	53	2	53	1	54	1	54	2	53.938	1.5
5	East Anglia	52	3	52	2	53	2	53	3	52.625	2.625
6	Southern Array	50	0	50	–1	51	–1	51	0	50.625	–0.188
7	West Isle of Wight	50	–1	50	–2	51	–2	51	–1	50.375	–1.75
8	Bristol Channel	51	–4	51	–5	52	–5	52	–4	51.375	–4.5
9	Irish Sea	53	–4	53	–5	54	–5	54	–4	53.75	–4.25

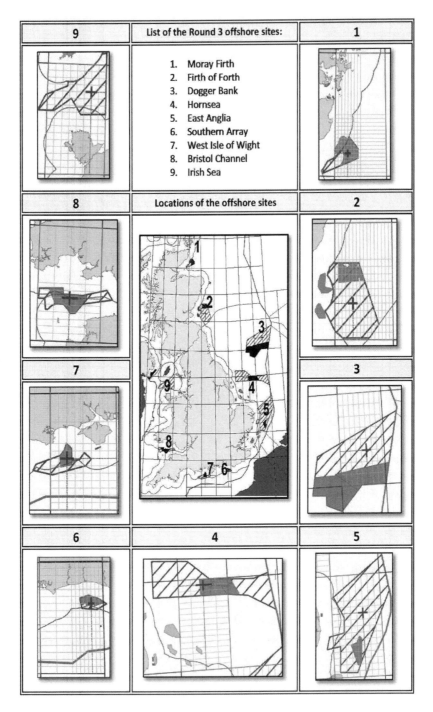

Figure 1.1 UK map with the geographical location of the Round 3 sites

The interpolation point of each Round 3 site is also shown.

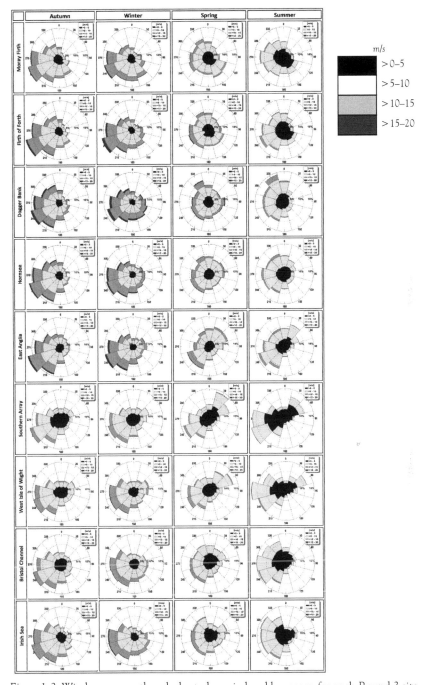

Figure 1.2 Wind roses over the whole study period and by season for each Round 3 site

The diurnal behaviour of the wind direction is shown in Figure 1.3. This time, the data are presented as a frequency distribution to enable an easier comparison of the wind distribution inferred from the six-hourly reanalysis data.

of the sites show significant differences in the sectors from which the wind blows most frequently. East Anglia, Southern Array and West Isle of Wight show a tendency to an increase in north-easterly winds during the spring and summer months. Dogger Bank and to some extent Hornsea see an increase in northerly winds during the spring and summer. For Bristol Channel and the Irish Sea site, shelter due to the land reduces variability though there is a slight tendency for more north-westerly winds. The far northerly Scottish sites of Moray Firth and the Firth of Forth show the least variation in direction.

The diurnal behaviour of the wind direction is shown in Figure 1.3. This time, the data are presented as a frequency distribution to enable an easier comparison of the wind distribution inferred from the six-hourly reanalysis data.

It can be seen that although there is some variation in wind rose with time of day, this variation is not large across the sites. The Bristol Channel site shows some evidence for a funnelling of the wind along the Severn Estuary during the later afternoon driven by a sea breeze. East Anglia shows a tendency to more north-north-easterly winds in the later afternoon which is probably due to a sea-breeze effect. Southern Array shows a tendency to stronger channelling along the north-east to south-west axis during later afternoon indicating a combination of

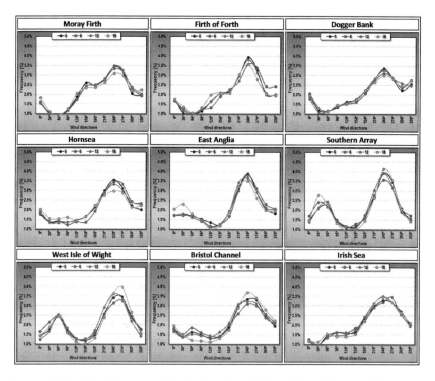

Figure 1.3 Directional frequency distributions for each Round 3 site by time of day

The legend gives the time of day in the form: 0 = 00Z, 6 = 06Z, 12 = 12Z, 18 = 18Z.

orographic channelling and a sea breeze. Dogger Bank shows virtually no variation which might be expected as it is furthest from land. Interestingly, Hornsea shows some variation despite also being one of the furthest offshore sites. The Irish Sea site and the Scottish sites show no strong influence due to the UK or Irish land masses.

Wind speed

Figure 1.4 shows the variation in 10 m wind diurnally and seasonally for the Round 3 sites. Diurnal variation is relatively small and all sites display a peak daily wind speed around midday with the exception of Hornsea and East Anglia where the peak is in the evening. All sites show significant seasonal variation with highest winds in winter and lowest in summer as noted previously.

Figure 1.5(a) shows the average fractional variation in daily wind speeds for the Round 3 sites as a function of distance from the coast. It can be seen that there is a

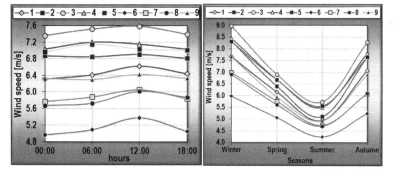

Figure 1.4 Diurnal (left) and seasonal (right) variation in the 10 m mean wind speed for the Round 3 sites

ID numbers: 1 = Moray Firth, 2 = Firth of Forth, 3 = Dogger Bank, 4 = Hornsea, 5 = East Anglia, 6 = Southern Array, 7 = West Isle of Wight, 8 = Bristol Channel, 9 = Irish Sea.

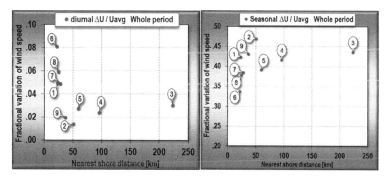

Figure 1.5 (a) Diurnal and (b) seasonal variation as a fraction of average wind speed for each of the Round 3 sites

Key to the sites as in Figure 1.4.

strong dependence on proximity to the land as might be expected, with variation up to 8 per cent for the closest site. Beyond around 30 km, there is little change in diurnal variation which tends to around 1–3 per cent. Figure 1.5(b) shows the average fractional variation in seasonal wind speeds for the Round 3 sites as a function of distance to the coast. This shows almost the opposite behaviour to the diurnal variation with the least variation seen for sites closest to the coast (minimum of 33 per cent) and greatest for those further away (41–46 per cent).

Figure 1.6 shows the distributions of 10 m wind speed for each of the sites with a Weibull distribution fitted using a Maximum Likelihood Estimator (MLE) methodology (Flygare 1985). It can be seen that a Weibull function fits each of the distributions well. Table 1.2 shows the Weibull parameters fitted to the 10 m wind distributions for the Round 3 sites as a function of season. Again, this confirms the expected seasonal variation in the mean wind speed, but also shows less variability in the autumn/winter seasons. Although k is seen to be significantly greater than 2 for all seasons indicating less variable winds than onshore, this should be viewed with caution as the values are based on six-hourly values so variation will be inherently lower than hourly or ten-minute wind speeds frequently available on land.

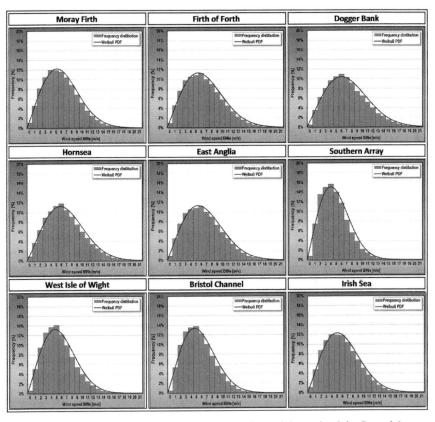

Figure 1.6 Distribution of the long-term 10 m wind speed for each of the Round 3 sites with a Weibull distribution fit superimposed

Table 1.2 Fitted Weibull parameters for the 10m wind speed distribution as a function of season for the Round 3 sites

	Moray Firth		Firth of Forth		Dogger Bank		Hornsea		East Anglia	
	k	c[m/s]	k	c[m/s]	k	c[m/s]	k	c[m/s]	k	c[m/s]
Winter	2.38	8.68	2.42	9.42	2.36	10.2	2.39	9.55	2.42	9.42
Spring	2.15	6.98	2.10	7.26	2.01	7.87	2.26	7.57	2.10	7.26
Summer	2.18	5.61	2.08	5.81	2.19	6.54	2.24	6.24	2.08	5.81
Autumn	2.35	8.01	2.34	8.66	2.32	9.40	2.32	8.84	2.34	8.66

	Southern Array		West Isle of Wight		Bristol Channel		Irish Sea	
	k	c[m/s]	k	c[m/s]	k	c[m/s]	k	c[m/s]
Winter	2.21	6.81	2.30	7.98	2.35	7.90	2.42	8.66
Spring	2.18	5.79	2.22	6.58	2.18	6.44	2.14	6.75
Summer	2.24	4.88	2.25	5.44	2.18	5.41	2.12	5.60
Autumn	2.05	5.99	2.14	6.95	2.20	6.98	2.29	7.82

Wind resource assessment case studies

Introduction

The application of a mesoscale model to the prediction of offshore wind resource is detailed in this section. The research involved the use of the Weather Research and Forecasting (WRF) mesoscale model (Janjic 2003, Skamarock 2008) to assess the wind conditions at selected sites in UK offshore waters. Specifically, the Advance Research WRF model core (ARW) was used in this work. The accuracy of the model was assessed in a number of ways: 1) through application of several planetary boundary layer (PBL) schemes, both individually and as an ensemble; 2) through the use of time-step ensembles; 3) by the use of different timescale filters; 4) through the use of model 'nudging' using nearby observations. Each model run had its boundary conditions set using output from the National Centers for Environmental Prediction (NCEP) Climate Forecast System Reanalysis (CFSR) (Saha 2010). This research was therefore concerned with how well a mesoscale model could downscale global forecast analysis data. A comparison was made between model output and observations from meteorological masts at Scroby Sands off the east coast of the UK and two masts at Shell Flats off the north-west coast. Model performance is assessed in terms of ability to predict wind speed and atmospheric stability. Recommendations were made in terms of how best to use the model for offshore wind resource prediction. Finally, a projection was made of the wind conditions at a future potential offshore wind farm site in the UK Round 3 Dogger Bank development zone. The variation in synoptic wind conditions across a large hypothetical 1.2 GW wind farm in this area were also assessed including maximum expected wind speed and wind direction differences across the wind farm.

Background

WRF has become widely utilised in the atmospheric sciences research field. It has been applied to a full spectrum of investigations which include high resolution simulations, e.g. (Litta 2008), which are relevant to wind resource assessment.

A number of studies have used WRF for offshore wind resource assessment. Tastula (2012) reports an investigation undertaken into the performance of WRF compared with the ERA-Interim reanalysis product which was also used as initialisation and boundary data for the model run. The performance of the model was studied in the boundary layer which is of particular relevance to this study. Findings showed the model to offer a higher level of performance than the ERA-Interim reanalysis product for the vast majority of variables studied apart from surface pressure. However, this was attributed to the provision of buoy data which was incorporated into the ERA-Interim product but not the WRF model run. The US army have investigated the operational use of WRF, at a high resolution, e.g. 0.3–3 km, for the purposes of very short term forecasting and nowcasting applications (Dumais 2009). For some locations, the use of WRF to create an offshore wind resource assessment product has already been undertaken, e.g. Peña (2011) describes a wind atlas for the South Baltic region. Such an application was essential because of the lack of observational data to the south of the region, while output was validated at locations in the domain where observational series were available from Danish and German masts. WRF has the potential to perform well as a wind resource assessment tool and has already been applied in the production of a wind atlas, which makes the next step validating performance for use as a site assessment tool, both in a historical long-term context and short-term operational context. A review was produced (Zhao 2012) for a system which is operational in China whereby GFS forecast data are downscaled by WRF and passed through a Kalman filter for the purpose of day ahead forecasting. It was found that the system performed with an acceptable level of error (16.47 per cent normalised root mean squared error (RMSE)). Some traits of the model itself and setup options have been identified which should be considered when undertaking a wind resource study. The limit to the potential performance of the model is somewhat constrained by computing resource. In order to optimise a model run, outright resolution is often compromised to achieve a quicker model runtime and reduced computational resource requirement. In theory, the higher the simulated resolution, the better model performance would be as more processes are able to be directly resolved. However, it was found (Gibbs 2011) that increasing resolution around the 4 km range yielded diminishing returns with respect to the subsequent extra requirement in computing resource and instead suggested utilising larger spatial domains and vertical resolution to try and improve resolution of the larger scale features. Operationally, WRF has been shown to possess a high surface wind speed bias, e.g. (Mass 2011, Jiménez 2012). Knowledge of such a bias can be beneficial, as it allows for possible systematic correction in future predictions. Such a bias, however, might cause problems in model simulations which involve a coastal interface.

Methodology

Sites

Two observational data series were used for validation in this work, namely Scroby Sands and Shell Flats (Figure 1.7). Ten-minute averaged data were collected at both sites. At Scroby Sands, temperature, wind speed and wind direction were measured at 33 m and 51 m, from 1995 to 2000. There were some periods of missing data and this had an influence on model run periods. Two masts were erected at Shell Flats. At Mast 1 wind speed, wind direction, temperature, relative humidity, pressure, rainfall and solar radiation with instruments were recorded at 12 m, 20 m, 30 m, 50 m, 70 m, 80 m and 82 m above highest astronomical tide (HAT). Observations at Mast 2 were made at 12 m, 20 m, 30 m, 40 m and 52 m above HAT.

Observational data from two onshore meteorological stations were also used in this study, namely one at Hemsby and one at Squires Gate (Figure 1.7).

The meteorological conditions at a 'hypothetical' site in one of the UK's Round 3 offshore wind farm development zones (Dogger Bank) were simulated to investigate the expected variability in wind speed and direction as well as atmospheric stability. This is also indicated in Figure 1.7.

Figure 1.7 Locations of the three offshore and two onshore sites used in this study (offshore sites are marked as open circles and onshore sites as closed circles)

Model set-up

To run the model, two high performance clusters were used, namely Loughborough University's Hydra cluster which is comprised of 161 compute nodes, each having two six-core Intel Westmere Xeon X5650 CPUs and 24 GB of memory; and the UK Engineering and Physical Sciences Research Council (EPSRC) national supercomputing facility HECToR (High-End Computing Terascale Resource). HECToR has 2816 compute nodes, each with two 16-core AMD Opteron 2.3 GHz Interlagos processors and 32 Gb of memory. Aside from significant processor power, HECToR possesses advanced data communication hardware such that each 16-core socket is coupled with a Cray Gemini routing and communications chip which translates to data latency between two nodes of around 1–1.5μs. HECToR runs Linux and is available with many selectable modules and compilers, for example, gfortran, PGI, Intel and Cray. Ideally, HECToR would have been used for all runs, but the run-time allocation on this machine is limited due to demand.

   Three levels of nested domain were used for the ARW runs as shown in Figure 1.8 for Scroby Sands. A similar nesting configuration was centred on Shell Flats

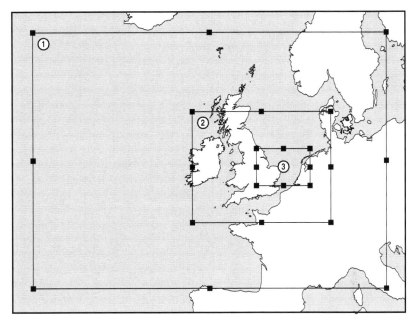

Figure 1.8 Three nested domains used for the ARW runs with resolutions of 18 km (outer), 6 km (middle) and 2 km (inner). The set-up for Scroby Sands is shown here

and the Round 3 site. Nests were offset to give more space for the model to simulate features originating over the Atlantic, where many weather systems which influence the UK originate. The 0.5° CFSR reanalysis product was used to initialise the model, which equated to a grid spacing of around 55 km. During the testing phase, breaches of the CFL (Courant Friedrichs Levy) criterion in the vertical plane were causing the model run to stop. The number of vertical levels was reduced to 50 vertical model levels which resolved the issue of numerical stability. Vertical levels were fairly evenly distributed apart from close to the surface where more levels were concentrated to improve resolution in the PBL. Fifteen levels were located below 500 m at 0, 20, 40, 65, 90, 110, 130, 150, 170, 190, 230, 270, 330, 405 and 490 m.

The dynamical options used for the mesoscale model runs are given in Table 1.3. Two PBL schemes were used: for initial testing the Mellor–Yamada–Janjic (MYJ) scheme (Janjic 2001) and later the Yonsei University (YSU) scheme (Hong 2006).

Scroby Sands was studied first for a relatively limited number of test cases in order to 'benchmark' the model. In this case, 34 periods were simulated over a year for model predictions extending from $t + 0$ h to $t + 90$ h. In the case of Shell Flats, a much larger number of simulations was run to simulate, as far as possible, a continuous period over 18 months. In this case, model runs were undertaken such that 489 days were simulated where each run extended from $t + 0$ h up to $t + 90$ h which was the longest look-ahead time for which model predictions were assumed to provide reasonable predictions, i.e. RMSE errors were not showing a significant increase.

Table 1.3 Dynamical parameters used in the ARW mesoscale modelling for Scroby Sands (SS), Shell Flats (SF) and the Round 3 site (R3)

Model parameter	Set-up
Vertical model levels	50 (SS), 40 (SF, R3)
Nesting feedback	On
PBL scheme	Mellor–Yamada–Janjic (MYJ) for SS and Yonsei University (YSU) for SF/R3
Cumulus scheme	Betts–Miller–Janjic
Radiation scheme – Long wave	GFDL
Radiation scheme – Short wave	GFDL
Microphysics option	Ferrier (new Eta) microphysics
Surface layer physics	Monin–Obukhov (Janjic)
Land surface option	Unified Noah land-surface model

For the hypothetical Round 3 offshore wind farm site, simulations were undertaken to cover a period of one continuous year, once again from forecasts up to $t + 90$ h.

Model nudging

Observational nudging is an objective analysis technique whereby an observational series is assimilated into the model input data. While large-scale model input data are convenient due to global coverage, homogeneous levels and a wide range of variables, coarse resolution might not be exactly representative of conditions at, or near, a site of interest. Nudging the model input using objective analysis is intended to improve the first guess of particular variables at, or close to, a particular site. Given that WRF is updated for the duration of a model run by input and boundary files, nudging is performed throughout the whole run. Nudging was tried using data from Hemsby for the Scroby Sands assessment and in the case of Shell Flats, Mast 1 was used to nudge for simulations at Mast 2. There was the possibility to nudge using data from Squires Gate, but this is not reported in this chapter. Only wind speed data were used to nudge the model as this was felt most relevant, though future investigations may use other variables. Nudged and non-nudged simulations were compared.

Results

Scroby Sands

It was clear that due to the limited temporal resolution of the model, that WRF would be unlikely to capture the observed variability at 10-minute intervals. Indeed, performance problems at short temporal scales were found (Nunalee 2014), where variation in model output appeared dampened in comparison to observations. With the innermost model domain resolution being 2 km, the smallest features which can be expected to be well resolved are around 14 km in size. Below 14 km, the model is able to account for atmospheric features to an extent, but does so through parameterisation schemes, specifically the planetary boundary layer scheme. Given that the temporal resolution of the runs is 10-minutely, it is unlikely that model performance will be best at simulating high frequency change as the size of atmospheric features responsible for change in wind speed on such a timescale is smaller than the directly resolved scale of the model. In order to investigate model performance on longer timescales at which atmospheric features are directly resolved, temporal filtering was performed on model runs and concurrent observations. Initially, an unweighted moving average filter was applied to the 10-minutely model output and Scroby Sands observations at intervals of 3, 9 and 17 time steps which corresponded to 30, 90 and 170 minute periods. Subsequently a low-pass Butterworth filter was also developed to filter out frequencies below 60, 180 and 360 minutes.

Table 1.4 summarises the average results from all 34 runs. RMSE and Pearson correlation coefficient between model and observed values is shown. As a benchmark, the wind speed at Hemsby is used as a simple predictor of the wind speed at Scroby Sands and the correlation coefficient calculated. A clear improvement is evident from both of the filtering processes. While filtering should intuitively reduce the variation in a series, the model output must still exhibit similar characteristics to the observations in order for the correlation to improve. Results are improved for the three-hour time increment by a greater margin using the moving average filter over the Butterworth filter and the performance gap compared with the simple Hemsby prediction is reduced.

These results confirm the value of using the model when applied to simulate features of appropriate scale. When done so, model output would seem to a good substitute for measurements at a nearby coastal meteorological station at least in the case of an offshore site relatively close to land.

Table 1.4 Pearson correlation coefficient and RMSE comparing WRF model predictions and observed data at Scroby Sands with various temporal filters. Hemsby is included as a benchmark predictor

	Hemsby	Hemsby MA	WRF	WRF MA
Effective temporal resolution (minutes)	60	180	10	170
Correlation	0.746	0.785	0.64	0.72
RMSE (m/s)			2.2	1.9
WRF Butterworth filtered				
Effective temporal resolution (minutes)	60	180		360
Correlation	0.662	0.698		0.733
RMSE (m/s)	2.1	2.0		1.8

Shell Flats

Table 1.5 summarises the RMSE and correlation coefficients for model predictions when compared with measurements for Mast 2 at Shell Flats. In this case, Squires Gate and Shell Flats Mast 1 are included as predictors and the correlation coefficients reported. Various Butterworth filtered predictions are compared on timescales between 10 minutes and 360 minutes.

It can be seen in this case that the RMSE is similar as for the Scroby Sands prediction with a reduction with increasing timetable. However, the correlation is significantly higher. In addition, the wind speed data at the onshore site at Squires Gate shows a rather lower correlation than in the case of Hemsby and Scroby Sands. The correlation is lower than that for the WRF model predicted wind speed. The wind speed data from the Shell Flats Mast 1 shows a much higher correlation than Squires Gate and slightly higher than the WRF wind speed.

Model nudging

For two periods of a month (July and October 2003) at Shell Flats model runs were undertaken with observational nudging using wind speed only from Mast 1. Statistics for these periods can be found in Table 1.6 and Table 1.7 for July and October, respectively, where the 'raw' observations from Mast 1 are presented as a benchmark. July 2003 provided the first case study, where the correlation coefficient between observed and modelled wind speed was improved by the nudging process. Interestingly, the correlation coefficient between observed and simulated direction also improved, albeit marginally. RMSE of the nudged wind speed time series was also found to be lower than the non-nudged series. Similarly, RMSE for wind direction was again slightly improved by nudging the speed with the nudged direction RMSE value slightly lower than that of the non-nudged. October 2003 provided the second case study, in which the correlation coefficient for wind speed was marginally higher for the non-nudged run compared with the nudged run. Similarly, RMSE was marginally higher for the nudged run compared with the non-

Table 1.5 Pearson correlation coefficient and RMSE comparing WRF model predictions and observed data at Shell Flats Mast 2 with various temporal filters. Squires Gate and Shell Flats Mast 1 are included as predictors

	Squires Gate		Shell Flats Mast 1	
Effective temporal resolution (minutes)	60		10	
Correlation	0.59		0.94	
WRF Butterworth filtered				
Effective temporal resolution (minutes)	10	60	180	360
Correlation	0.856	0.865	0.883	0.901
RMSE (m/s)	2.1	2.1	1.9	1.7

Table 1.6 Statistics for the July simulation period showing the performance of WRF as a predictor of the wind speed at Mast 2, with and without nudging from Mast 1. Comparison is made with raw data from Mast 1 as a simple predictor. Heights are 40m above HAT

		Shell Flats Mast 1	*Nudged model (Model + Mast 1)*	*Non-nudged model*
Speed	Correlation coefficient	0.934	0.81	0.739
	RMSE (m/s)	1.2	2.1	2.6
Direction	Correlation coefficient	0.886	0.8	0.79
	RMSE (deg)	31.5	44.4	46.9

Table 1.7 Statistics for the October simulation period showing the performance of WRF as a predictor of the wind speed at Mast 2, with and without nudging from Mast 1. Comparison is made with raw data from Mast 1 as a simple predictor. Heights are 40m above HAT

		Shell Flats Mast 1	*Nudged model (Model + Mast 1)*	*Non-nudged model*
Speed	Correlation coefficient	0.919	0.888	0.889
	RMSE (m/s)	1.8	1.9	1.9
Direction	Correlation coefficient	0.644	0.65	0.622
	RMSE (deg)	50.6	53.6	56.1

nudged run. By contrast, a slight improvement in wind direction was observed, with a higher correlation coefficient and a lower RMSE for the nudged run.

From these results, it can be seen that nudging can sometimes improve predictions and sometimes does not, though this may be related to how good the correlation is initially.

PBL and PBL ensemble predictions

To assess the performance of different PBL schemes, 20 test periods were run for Scroby Sands. The results of this study are shown in Table 1.8 with the best performing schemes assessed in terms of highest correlation and lowest RMSE for the 20 cases. In Table 1.8, the individual PBL schemes are run with nudging using wind speed data from Hemsby, with the exception of one set of simulations using the MYJ scheme where nudging was not included. An equally weighted ensemble of all of the PBL schemes was also analysed. In general across the runs undertaken, statistical performance of the schemes is very similar. The MYNN and ACM2 schemes display the best average statistics, very close to those of the ensemble mean, and perform the best in the highest number of cases for the nudged PBL schemes.

Table 1.8 A comparison of PBL schemes in terms of WRF model performance at Scroby Sands

PBL scheme	Number of cases as top performer		Av. statistics	
	Corr.	RMSE	Corr. coeff.	RMSE (m/s)
MYJ	8	8	0.577	2.4
MYNN	12	8	0.602	2.4
ACM2	11	12	0.599	2.4
QNSE	3	5	0.551	2.5
MYJ (no nudging)	16	12	0.558	2.5
PBL ensemble	10	15	0.607	2.4

Formulation of the ACM2 PBL scheme suggests it should be a capable performer under unstable conditions, which might account for its level of relatively high performance compared with the other schemes as previous work (Argyle 2012) has suggested that unstable conditions persist at Scroby Sands for a large proportion of time. The remaining schemes, MYJ and QNSE are not especially poor performers, though the QNSE scheme does fare less well compared with the other schemes. The technical difference between the MYJ and MYNN schemes is in the formulation of the master mixing length scale, which might be the reason for the observed difference in performance in this study. In the MYJ scheme, the mixing length is a function of height, where in the MYNN scheme, turbulence, buoyancy and surface length scales are all used to form the mixing length scale, which all provide more detailed information regarding the turbulence present contributing to fluxes through the boundary layer. The QNSE scheme displaying the lowest performance is not so surprising, as it is tuned for stable conditions. Further work is required to identify the specific nature of the test cases, for example identifying if they were neutrally, stably or unstably stratified, which could feed into the development of a more 'intelligent' ensemble mean with appropriate weighting.

Time Offset Ensemble System (TOES)

As well as ensemble averages over all PBL schemes, an additional ensemble average was studied, namely the Time Offset Ensemble System. As each run was over 90 hours, there was the possibility of starting a new run at intervals during the 90 hours. In this case, the option of starting a run at 24 and 48 hours into the initial run was investigated for Scroby Sands. Ensemble averages were produced of the original, the next reinitialised run $t + 24$ hours later, and a third $t + 48$ hours after the original. For each of the three time offsets, all PBL schemes were run as reported in the previous section giving ensemble averages over time offset and PBL runs. Summary statistics after applying these methods to predicting the wind speed at Scroby Sands are shown in Table 1.9. This table also includes

Table 1.9 Summary statistics for the TOES methods applied to Scroby Sands

Comparison beginning		Corr. coeff.	RMSE (m/s)
t+24h	PBL ensemble	0.5591	2.6
	PBL/time offset ensemble	0.6003	2.4
t+48h	PBL ensemble	0.5862	2.4
	BPL/time offset ensemble	0.6374	2.2

performance statistics for the PBL ensemble for the corresponding run without time offset averaging. It is seen that the combination of PBL and time offset averaging improves the correlation and reduces the RMSE. This would suggest that TOES are valuable in increasing prediction accuracy and that earlier run information still adds value even when a run period is reinitialised.

Atmospheric stability

For Shell Flats Mast 2, the atmospheric stability was classified using the Bulk and Gradient Richardson number inferred from measured data. This was then mapped to Obukhov length L and classified as either neutral ($|L| > 1000$ m), very unstable ($-200 \leq L < 0$), unstable ($-1000 \leq L < -200$), stable ($0 > L \geq 200$) or very stable ($200 > L \geq 1000$). The Gradient Richardson number was inferred from temperature data at 12 m and 82 m and wind speed data from 10 m and 82 m. The Bulk Richardson number was calculated using the same temperature data, but only wind speed at 82 m. WRF model Bulk Richardson number data were produced based on temperature data output at 10 m and 50 m as well as wind speed data at 40 m. As noted above the YSU PBL was used.

Figure 1.9 summarises the stability statistics thus calculated. It can be seen that the observed Bulk and Gradient Richardson number metrics give quite a different picture in terms of the prevailing atmospheric stability. The observed Gradient Richardson number statistics agree broadly (Argyle 2012) with predominantly unstable conditions, whereas the observed Bulk Richardson number suggests a more symmetrical spread, with fewer very stable or very unstable conditions. The reasons for this may be due to the way the two metrics are calculated; the gradient method can produce large values of the Richardson number when the wind speed values at the two heights are very close together. The Bulk method is relatively immune to this, but the mapping of Bulk Richardson number to Obukhov length is more tenuous. The modelled Bulk Richardson number shows a reasonable level of agreement with observations though there are a lower number of neutral conditions and a tendency to predict more stable conditions.

Figure 1.10 shows the stability statistics this time by direction sector. It can be seen that there are a larger proportion of unstable conditions when the wind blows from the north and more stable from the south reflecting the fact that colder northerly air overlying warmer water will tend to promote unstable

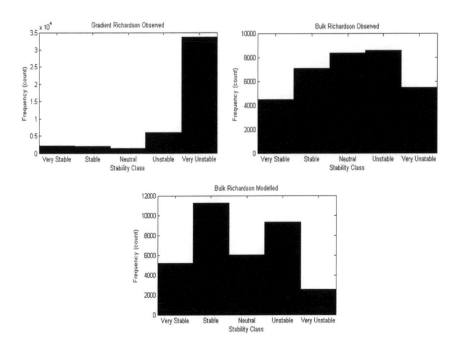

Figure 1.9 Observed and modelled surface layer atmospheric stability statistics for Shell Flats Mast 2

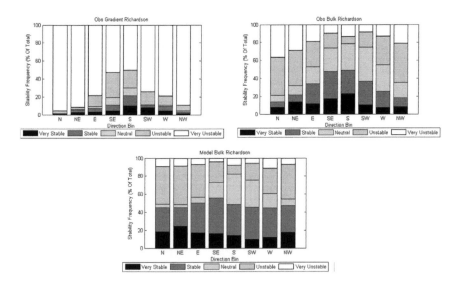

Figure 1.10 Observed and modelled surface layer atmospheric stability conditions at Shell Flats Mast 2 by direction

conditions whereas warmer southerly air overlying cooler water will tend to promote more stable conditions. The model simulates this trend quite well albeit with a tendency to predict more stable conditions as noted above.

Round 3 wind farm site

WRF simulations of the wind and stability conditions were carried out at a hypothetical site in the UK's Round 3 Dogger Bank Offshore Wind Farm Development Zone. Using a mesoscale model provides an opportunity to assess variability across a large area such as a large wind farm spanning over 20 km and in this case, data were extracted for a central point and for grid points spanning a hypothetical wind farm site of area 20 km × 20 km containing 1.2 GW of capacity. As for the main Shell Flats study, the YSU PBL scheme was used.

The overall distribution of wind speeds at the site suggests a Weibull-like distribution with scale parameter, C = 10.2 m/s and shape parameter, k = 2.13 at 90 m height. Figure 1.11 shows the projected wind rose for this site with a dominance of wind from the south moving clockwise round to the north-west where the highest wind speeds are projected.

Figure 1.12 shows the distribution of projected stability conditions at this site. The projection is for predominantly neutral conditions with some stable/very stable conditions observed, though this should be viewed with some caution as the model has a tendency to predict more stable conditions then observed, as noted above, at least using the YSU PBL scheme. Figure 1.13 shows the breakdown in stability by direction. The same tendency is noted as for Shell Flats with more stable conditions from the south, though this tendency is less pronounced for the Round 3 site.

Three months of simulations were also performed to gain an impression of the deviation in conditions seen across the farm. Due to the size of the Round 3 site,

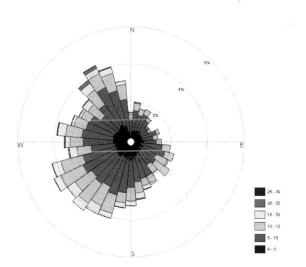

Figure 1.11 Projected wind rose for the Round 3 site. Scale is in units of m/s

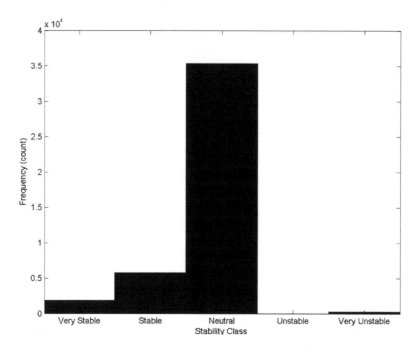

Figure 1.12 Stability distribution at the Round 3 site

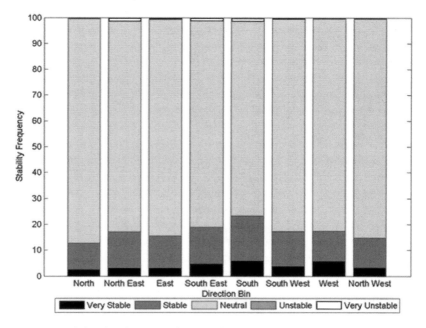

Figure 1.13 Stability distribution at the Round 3 site by direction

it is entirely possible that turbines at opposite extremities of the farm might be subject to different weather systems at the same time.

Figure 1.14 is an example of one occasion where wind direction was very variable over the extent of the wind farm area with an extreme difference of over 50 degrees. This could have significant implications for the overall performance of the wind farm due to wake interactions and highlights the importance of considering synoptic as well as smaller scale turbulent wake meandering and Coriolis effects when considering overall farm performance for such large potential sites.

Wind turbine wake effects

This section details the application of a computational fluid dynamics (CFD) model for the prediction of wake effects in large offshore wind farm arrays. The model used was the Ansys CFX code with the Windmodeller front-end for defining the initial set-up and post-processing the results. The CFX model was run in two configurations: the steady-state Reynolds Averaged Navier–Stokes (RANS) configuration and the Unsteady Reynolds Averaged Navier–Stokes (URANS) configuration. Two different types of turbulence models were tested, namely the k-ε model and Menter's Shear Stress Transport (SST) model which uses a k-ω model near the wall and a k-ε model away from the wall. The domain and mesh were not changed between simulations, ensuring an effective blind test. Turbines were simulated as actuator discs acting as momentum sinks whose thrust depends on the upstream wind speed according to the turbine thrust curve. Results were compared against production data from the Danish offshore wind farms Horns Rev and Nysted and against the UK offshore wind farms Robin Rigg and Scroby Sands whose locations are shown in Figure 1.15.

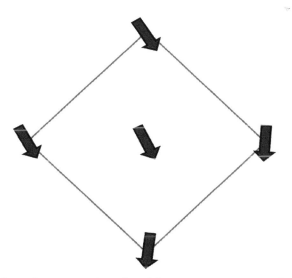

Figure 1.14 Schematic representation of wind direction variation across the Round 3 site on a certain day during the three-month simulation period

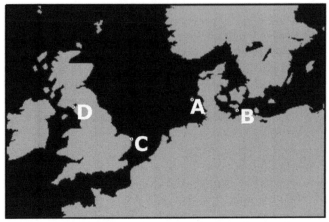

Figure 1.15 Locations of offshore wind farms for which production data were available for this work

A = Horns Rev, B = Nysted, C = Scroby Sands, D = Robin Rigg

Every farm location has a different environment. Despite the surface roughness length z_0 effectively remaining constant between offshore locations, conditions on the sea bed, local marine activities, distance from shore as well as conditions of the coastal area where the export cables make landfall can all play a part in dictating farm layout. This can be seen most clearly in Figure 1.16 with the Scroby Sands layout, where gaps and extra turbines in an otherwise regular pattern of three columns of turbines betray locations of sea bed unsuitable for turbine foundations. The Robin Rigg layout by comparison was constrained by shipping lanes and water depth. The Danish farms are regular in a parallelogram layout but with different orientation and separations. On account of each farm having differing layouts, conducting typical 'down the line' assessments are not always possible. This can be viewed as a problem for simulation validation as cross-farm comparison is harder, but it can also be viewed positively since Figure 1.16 shows offshore farms are not always in regular arrays and averaging power production across numerous turbines may hide unusual or unexpected events. Within these four farms there are three types of turbine design with varying rotor diameters and thrust curves. This further complicates analysis which can no longer be based on turbine spacing and meteorological conditions alone. As ever, there is always a need for more farms to contribute data towards model validation.

Each of the four farms shown in Figure 1.16 has at least one location where meteorological measurements are taken. Information about what is measured at each location is shown in Table 1.10. Note that not all of the met masts were fully maintained after farm construction was complete and therefore there are often periods of missing or erroneous data.

In addition to each wind farm being unique in its turbine layout, the actual turbines installed are different in each farm as well. This is mainly due to the

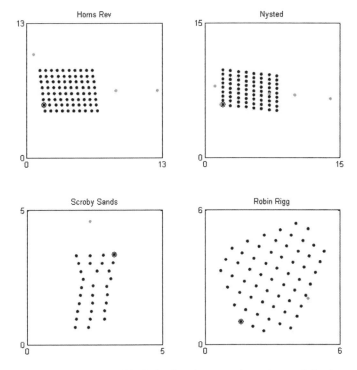

Figure 1.16 Layout of turbines (black dots) and meteorological masts/lidar (grey dots) at each of the four offshore wind farms which supplied production data for this work

The 'key turbine' used as a reference for free-stream wind direction and speed is highlighted by a ringed dot. All dots are the same size and do not represent the scale of the object they symbolise. Objects are located at the centre of a dot and distances between objects are indicated by the scale (in km) on the diagram edges. In all four plots, North is to the top in each case.

variations in farm age, although local planning regulations also play a part. This can be seen in Table 1.11 and Figure 1.17 by comparing the Horns Rev and Scroby Sands farms where both use the same turbine model although the British farm has a lower hub height to reduce visibility from shore. The Robin Rigg farm by comparison is made up from a more recent and thus larger model designed by the same company.

As large offshore wind farms by definition have many components that may require maintenance at any particular time, causing an individual or group of turbines to remain stationary, it is a rare event when all turbines are fully operational whilst the atmospheric conditions are also within boundaries set by a case study. Therefore, at the three largest farms, measured values from each turbine were only considered when every machine along that line of turbines was fully operational. Then these filtered values were averaged across turbine lines so that each of the free-stream turbine outputs were averaged together, all the output values from the second turbines in line were averaged together, and so on. Whilst this means stationary turbines in neighbouring lines may not be contributing

Table 1.10 Comparison of met masts associated with wind farms and their installed instrumentation. Additional instruments (such as rain gauges) are present at some masts, although these have been left out of this table as their measurements were not used as part of this work

| | Horns Rev | | | Scroby Sands | Robin Rigg Lidar Platform |
	M2	M6	M7		
Nationality	Danish			British	British
Water body	North Sea			North Sea	Irish Sea
Mast height (m)	62	70	70	51	10 (platform)
Heights (m) of cup anemometers	15, 30 45, 72	20, 30, 40, 50, 60, 70	20, 30, 40, 50, 60, 70	33, 51	10, 14, 58, 102, 114 (lidar)
Heights (m) of wind vanes	28, 43, 60	28, 68	28, 68	33, 51	10, 14, 58, 102, 114 (lidar)
Heights (m) of thermometers	13, 55	16, 64	16, 64	19, 47	10 (platform)
Heights (m) of hygrometers	13	N/A	N/A	N/A	10 (platform)
Heights (m) of barometers	55	N/A	N/A	14	10 (platform)
Depth (m) of sea thermometer	4	4	4	N/A	N/A

| | Nysted | | | | |
	Mast1	Mast2	Mast3	Mast4	Mast5
Nationality	Danish				
Water body	Baltic Sea				
Mast height (m)	68	68	68	68	25
Heights (m) of cup anemometers	10, 25, 40, 55, 65, 68	10, 25, 40, 55, 65, 68	10, 25, 40, 55, 65, 68	10, 25, 40, 55, 68	10, 25
Heights (m) of wind vanes	68	68	68	68	N/A
Heights (m) of thermometers	10, 65	10, 65	10, 65	10	10
Heights (m) of hygrometers	10	N/A	N/A	N/A	N/A
Heights (m) of barometers	10	10	10	N/A	10
Depth (m) of sea thermometer	N/A	2	2	N/A	N/A

Table 1.11 Comparison of wind turbines installed at each of the four offshore wind farms

	Horns Rev	Nysted	Scroby Sands	Robin Rigg
Model	Vestas V80	Bonus 2.3	Vestas V80	Vestas V90
Rated power	2MW	2.3MW	2MW	3MW
Hub height	70m	68.8m	60m	80m
Diameter	80m	82.4m	80m	90m
Cut-in wind speed	4ms^{-1}	3ms^{-1}	4ms^{-1}	3.5ms^{-1}
Cut-out wind speed	25ms^{-1}	25ms^{-1}	25ms^{-1}	25ms^{-1}
No. of turbines	80	72	30	60

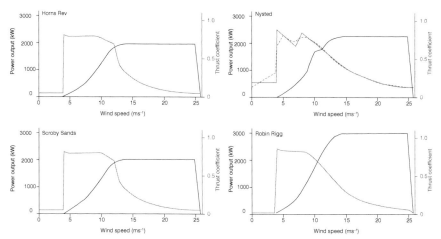

Figure 1.17 Plots of wind turbine power curves (dark solid line) and their corresponding thrust 1coefficients (light solid line) according to wind speed

Since Windmodeller struggled to converge any simulation using the sharp variations in thrust coefficient for turbines at the Nysted wind farm, an alternative, smoothed curve (dashed line in Figure 1.17) was used as an acceptable substitute.

to losses from horizontal wake expansion, the number of suitable validation events was significantly increased. To assist with clarity for the staggered Robin Rigg layout, Figure 1.18 shows how the definition of each turbine position in a line is calculated for this work. As each individual wind turbine has a blocking effect on wind flow and air at low speeds may be considered incompressible, regions between turbines often report flow velocity increases above the free-stream value, known as the venturi effect. The inclusion of some venturi affected turbines within the free-stream category should not bias the work significantly as they are not directly within the wake of another turbine and are consistent for both simulation and validation datasets. They may however be a topic for future investigation, to discover whether such venturi effects alter the rate of

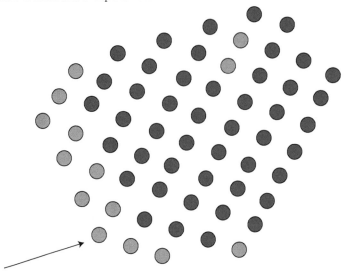

Figure 1.18 Layout of turbines at the Robin Rigg offshore wind farm

Highlighted (light grey top right) are turbines B2 and C3 which had significantly fewer events suitable for validation purposes and so were excluded from the analysis along with the turbines downstream of them. Highlighted (light grey bottom and left) are the free-stream turbines for case studies with wind direction 249° (indicated by arrow) showing that whilst they are not directly in the wake of an upstream machine, at least three turbines are likely to be affected by venturi processes as air flows between two other turbines. There are also five turbines which may be partially affected in this way as they are located diagonally behind at least one other machine.

wake dissipation within large offshore farms and whether such findings can be used to develop more productive wind farms.

A major assumption of this work is that any CFD model suitable for mass deployment throughout the wind industry must both combine the flexibility to adjust simulation parameters by experienced users and give acceptable results when used as a 'black box' technique by less experienced users or developers without time to conduct calibration tests. For example the mesh resolution affects both the simulation accuracy and the computational cost. Windmodeller's default setting (before automated mesh refinement takes place) is for a 100 m horizontal mesh within the centre block and an average vertical resolution of 80 m throughout a 2 km deep domain. Although this seems large, each mesh layer is defaulted to 15 per cent deeper than the one below it. Therefore, at an 80 m hub height this would result in a mesh roughly 16 m deep. After automated re-meshing takes place, the resolution around each turbine is considerably finer, enabling it to model the blockage effect and shear generated turbulence generated by the disc. The sensitivity study conducted by Carney (2012) systematically varied mesh input parameters in isolation of others; finding solution accuracy and cost were most sensitive to horizontal resolution and the domain radius, although trends were dependent on wind direction relative to the mesh angle. Their recommendations for domain depth were just less than 1 km with initial horizontal and average vertical mesh resolutions of 0.4 and 0.35 times the rotor diameter (D) respectively, although having only conducted tests on one farm it is uncertain whether these are site specific, if the scales depend on the size of D or are general RANS rules of thumb. When comparing four CFD models, Brower (2013) states that spatial resolution must be roughly 50 m or better which roughly correlates to that suggested by Carney (2012).

Each of the benchmark simulations were conducted using the same initial mesh geometries although the automated process of re-meshing around turbines and the actuator discs (ADs) positioned facing the flow may cause each individual case study to have slightly different mesh qualities. However, they are all initiated as a 1 km deep cylinder with a 10 km radius. Horizontal resolution was set at a generic 50 m while average vertical resolution measured 45 m. To account for variability in production data wind direction, a bin size of five degrees was allotted to each case, with the modelled production being the average of three runs, one in the centre and one at each direction extreme. Table 1.12 lists the seven case studies along with the symbols plotted in the figures.

Figure 1.19 shows the measured production levels for individual turbines to be used as validation data for each case study, except for the top left plot corresponding to HR270, where each point represents the average measured production of the central six rows unaffected by the farm edge. How each turbine number in Figure 1.19 relates to its position in the farm according to Figure 1.16 is not immediately clear, especially for the farms Nysted and Robin Rigg, which would also benefit from similar treatment to help display their wake-losses as a function of turbine position within a row. However, displaying each turbine separately reveals some interesting phenomena such as the 'power cascade' in case N278; this indicates that whilst each column deeper in the farm generates less power than the previous column (sometimes referred to as the deep-array effect), not all turbines in the column generate the same amount of power, despite the column being orientated perpendicular to the wind direction. Indeed, there appears to be a clear north–south gradient where turbines closer to the north farm edge generate more power than those closer to the south in the same column, and often generate more than those close to the south edge of the previous column as well. Since the wind

Table 1.12 List of seven validation case studies for benchmarking

Wind farm	Wind direction	Case name	Symbol
Horns Rev	270 ± 2.5°	HR270	Solid
Nysted	180 ± 2.5°	N180	Open
Nysted	278 ± 2.5°	N278	Solid
Scroby Sands	77 ± 2.5°	SS77	Open
Scroby Sands	90 ± 2.5°	SS90	Solid
Robin Rigg	204 ± 2.5°	RR204	Solid
Robin Rigg	249 ± 2.5°	RR249	Open

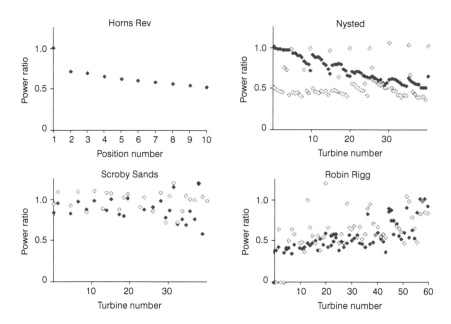

Figure 1.19 Plots showing measured power generated by each turbine normalised by the power generated by a turbine in the free-stream

Top left shows production data averaged across turbines in the six central rows at Horns Rev for case HR270 with the position number referring to each turbine's position in the 'down the line' row with position one being the free-stream turbine. Top right shows production data from Nysted case N180 (open symbols) and N278 (solid symbols). Bottom left shows production data from Scroby Sands case SS77 (open symbols) and SS90 (solid symbols). Bottom right shows production data from Robin Rigg case RR204 (solid symbols) and RR249 (open symbols).

direction is assumed to be uniform distributed with the average in this case study being 278°, it must be concluded that the northern turbines were not overly benefiting from biased event frequencies within the directional bin any more than the southern turbines. It is possible that since the northern turbines are closer to shore roughly 10 km away, they may be benefitting from higher levels of background turbulence and thus faster wake dissipation rates. However, since there is some evidence of a similar cascade effect in the eastern turbines for case N180, coastal proximity may not be the only cause, with Coriolis turning possibly also being significant across the length scales of an offshore wind farm.

From the numbering system used for Robin Rigg in Figure 1.19, it is difficult to observe any patterns as a result of 'down the line' sampling. However there is a general trend for both RR204 and RR249 that turbines located towards the backs of rows, so for these examples located to the north-east with lower turbine numbers, generate less power than those towards the front of rows, as would be expected. Due to missing data or turbine downtime, there were no suitable generation events for either turbine B2 or C3 (numbered 4 and 10 in Figure 1.19). Therefore, their points and the turbines directly behind them (turbines A1 numbered 1 for case RR204 and turbines A2 and B3 numbered 2 and 5 for case RR249) are shown as zero and will not be included in the analysis.

Figure 1.20 shows turbine power output as averaged by its position in a row of turbines. Scroby Sands is a relatively small farm with only 30 turbines (numbered from 1 to 38, with a number of originally planned machines not finally installed), which are not in strict lines characterised by the other three farms. Therefore, it is difficult to average the output of turbines by row position for this site. Therefore, the output of Scroby Sands is considered at a farm level in this study, whilst ignoring turbines 6, 10 and 21 as outliers resulting from a very small data sample and turbine downtime. However, production from the downstream turbines 5 and 9 are not ignored as they do not appear to be affected and it is therefore assumed the problems with turbines 6 and 10 are related to data collection rather than turbine downtime. Turbine 21 consistently registered SCADA data wind speeds and directions significantly different to other turbines and so was ignored with assumed data collection errors.

From Figure 1.20, it can be seen that four out of five large-farm case studies imply some level of deep-array effects. Case N180 is the exception, showing varying levels of generalised wake recovery between turbines; other than after the first and fourth turbine, each machine generates more power than the machine upstream of it. This is possibly a result of relatively close turbine proximity (5.8D

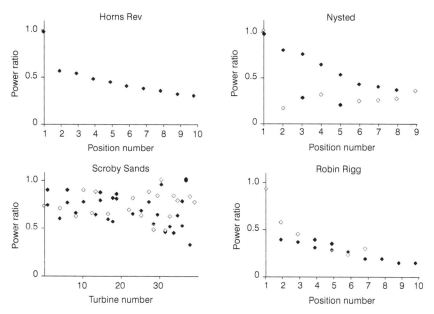

Figure 1.20 Plots of averaged power ratios through the farm

Edge effects are assumed to be negligible, considering the 5° direction bin and it dramatically simplifies calculations for Robin Rigg with its non-parallelogram boundary and two erroneous turbines. Position number refers to each turbine's position in the 'down the line' row with position one being the free-stream turbine. Due to farm layout, free-stream turbines for RR249 may experience some venturi effects as suggested by Figure 1.18.

separation), combined with a complex relationship between inlet wind speed and the thrust coefficient curve indicated by Figure 1.17. Case N278 shows the same turbines, for the same incident wind speed but different direction, generate more electricity per turbine when arranged with a larger distance between them (10.5D). However, the difference in generation between comparable turbine position numbers decreases further into the farm. This relation between turbine separation and productivity is again seen with the Robin Rigg data, although to a lesser degree. Case RR204 (separation 5.3D) has less initial wake loss than case RR249 (separation 4.3D), although the difference in turbine separation distance is less relevant further into the farm. Taken together, this suggests future farms may benefit from layouts with turbines separated by variable amounts depending on distance between turbine and the prevailing free-stream farm edge.

SST RANS

SST is the first turbulence model considered in this study. Figure 1.21 shows four plots comparing power generated at each wind farm against model output. Production data have not been filtered by stability category to help test the common assumption that simulations representing neutral stability are representative of average conditions. Incident hub height inflow is 8 ms^{-1} at the key turbine (as indicated in Figure 1.16) and modelled generated power has been normalised against measured power generated by this key turbine.

With the exception of N180, each plot in Figure 1.21 shows a slight over-prediction of power generation for any turbine in the free-stream, while they all overestimate wake losses in downstream turbines. For both Horns Rev and Nysted, the CFD model predicts a large initial drop in production between the first and second turbine with little additional loss through the farm. However, for Robin Rigg, the CFD model predicts lower initial losses with a sharp decline in generation with depth into the farm before levelling out to a near constant wake loss. Whilst the predicted wake losses in the top two graphs follow similar patterns, the difference in simulation results for the Robin Rigg farm may partially be a factor of turbine size; 50 per cent more rated power than either the of top farms leads to different levels of thrust and incident wake effects associated with each AD. Although a wind direction of 77° at Scroby Sands results in some wake losses (quite accurately modelled), the initial free-stream generation is systematically over-predicted. This over-prediction is worse for case SS90, with modelled free-stream generation roughly 50 per cent larger than measured. The modelled wake affected generation by contrast is nearly 50 per cent lower than measured values, so whilst the predicted overall farm generation may be close to reality, modelling of wakes and individual turbine behaviour is far from it.

k-ε RANS

The second turbulence model considered is the k-ε model as frequently utilised elsewhere, e.g. (Crasto 2013). Although it has been shown to be lacking in the

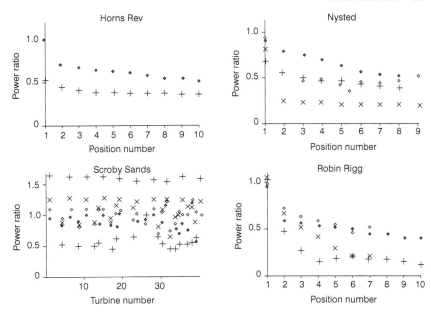

Figure 1.21 Results from SST simulations of the four farms marked as '+' and '×' symbols while diamond symbols denote the measured values for each farm

Top left shows Horns Rev for case HR270, top right shows Nysted case N180 (open and '×' symbols) and N278 (solid and '+' symbols), bottom left shows Scroby Sands case SS77 (open and '×' symbols) and SS90 (solid and '+' symbols) and bottom right shows Robin Rigg case RR204 (solid and '+' symbols) and RR249 (open and '×' symbols).

near wake (Rados 2009), where the application of AD theory is more problematic, the minimum separation within the four farms is greater than three diameters, the key distance indicated by Aubrun (2013) and Réthoré (2010). Figure 1.22 shows the results of simulations similar to those used to create Figure 1.21 except with the use of the k-ε turbulence model.

As for the SST results, for each simulation except N180, production at the free-stream turbine has been over-predicted, most significantly for SS90. With the Horns Rev and Nysted simulations, the k-ε turbulence model results fit the validation data better than the SST option, although it significantly under-predicts the wake losses at Robin Rigg. Whilst the k-ε model maintains reasonable accuracy for the first few wake-affected turbines in the top two farms, it appears not to match the deep-array effects seen most clearly in case N278. Prediction of further additional wake losses with depth into the farm does not occur, causing a plateau appearance whilst measured generation continues to drop for all three large farms. The variation in measured generation for case N180 implies a potential recovery from maximum wake losses whilst the simulation predicts a general plateau. Overall, modelled output using the k-ε turbulence model appears greater than that of the SST model as the power ratio values are larger, yet the

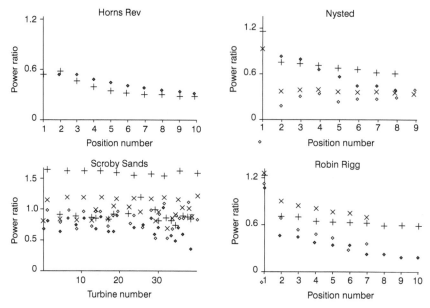

Figure 1.22 Results from initial k-ε simulations of the four farms

The symbols are the same as in Figure 1.21.

over-prediction at free-stream and deep-array turbines suggest the *k-ε* model is less accurate near flow boundaries and regions of strong adverse pressure gradients such as ADs. As indicated by Menter (2003), these boundary problems were the reason for the original development of the SST model, although it is not clear whether the SST model is better at modelling the deep-array effect than the .*k-ε*.

Modified *k-ε* RANS

The third turbulence model considered is a modified version of the *k-ε* model where one of the key model constants used in modelling the turbulent viscosity, C_μ, is altered from 0.09 to 0.03. This was done previously in Montavon (2011), and resulted in modelled power ratios comparable to those measured at the Horns Rev and North Hoyle farms, though the near-wake recovery process was suspected of being too fast, leading to the over-prediction of generated power at farms with closely spaced turbines. Figure 1.23 shows the results of simulations similar to those described above except with the use of the Modified *k-ε* turbulence model.

By comparing Figures 1.22 and 1.23 it can be seen that modifying the C_μ constant results in significantly better wake loss predictions for Robin Rigg, particularly case simulation RR249 except for a systematic power overestimation at every turbine. Measured production at position 6 is suspected of being abnormally low however, as it does not fit the smooth curve formed by either the other validation data points or the curve of simulation results. For case RR204,

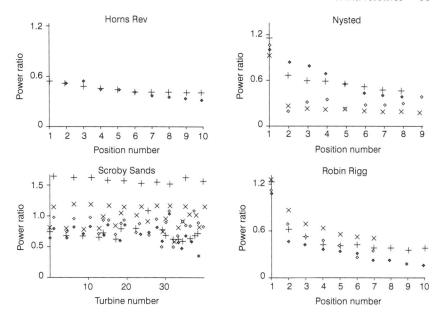

Figure 1.23 Results from the Modified k-ε simulations of the four farms

The symbols are the same as in Figure 1.21.

the Modified k-ε turbulence model provides a fair prediction throughout the early farm although fails to replicate the deep-array effect indiscernible for RR249 on account of farm design and performs better than both SST and the normal k-ε model at this farm. There is little significant improvement from modifying C_μ for either the Horns Rev or Nysted farms. Neither of the k-ε models captures the deep-array effect for HR270 and they just predict an effective wake-loss limit, although Modified k-ε returns higher total yield predictions than the basic k-ε model, and both predict greater yields than SST. Results from both simulations SS77 and SS90 give a better fit to the data for wake-affected turbines with the Modified k-ε model while N278 and N180 are inconclusive.

URANS

The final model used in this study is significantly different from the others in that the previous three were all RANS model runs computed in steady state with a disregard for thermal variations based on the neutral atmospheric assumption. The URANS model (Unsteady Reynolds-Averaged Navier–Stokes) by comparison is not run in steady state and is therefore more analogous to large eddy simulations (LES), although unlike LES, URANS still uses the cost saving averaging of turbulence values across all scales. Whilst there is additional cost to be paid in terms of computational time required for each simulation compared with basic RANS models, URANS benefits from the inclusion of a prognostic equation for potential temperature based on Montavon (1998). Although the atmospheric boundary layer (ABL) is often assumed to exhibit neutral stability, above the ABL there is a steady increase in potential temperature with height, resulting in a stably stratified layer. With the inclusion of the prognostic equation for potential temperature in URANS, it is possible to incorporate a layer at the top of the domain which varies in potential temperature with height. Thus, if a simulated air parcel moves into this region, its expansion and resultant cooling due to the lower pressure will force it to sink again, effectively 'capping' the simulation and helping to prevent uncontrolled model divergence. The strength and height at which this simulated free atmosphere region begins may be significant as it likely affects the expansion (and thus diffusion) of wakes behind each AD. Since the height of the ABL is known to vary significantly under different stability conditions, the height and strength of the marine free atmosphere thermal inversion should be the subject of future field investigation and is outside the scope of this work. For the purpose of URANS simulations which incorporate a free atmosphere thermal gradient, it shall be assumed that it matches the dry adiabatic lapse rate and begins at a height determined using work by Zilitinkevich (1996), whilst the simulated ABL itself remains neutrally stratified.

A further benefit of a transient simulation is that it allows for the inclusion of the Coriolis effect which not only controls the Ekman spiral (Carney 2012) but may also produce significant farm-edge effects over the large areas proposed for development in the North Sea (Seshadhri 2013). This effect is important as it can influence the flow direction and hence wakes losses through the extent of a large wind farm.

In order to observe the effects of the URANS components, results from four URANS simulations of case N278 are described in Table 1.13.

The results in Figure 1.24 show significant variation in simulated power ratios at each turbine location within case N278 dependent on which aspects of the full URANS model are included. For example, Option 3 and Option 4, which both include the Coriolis parameter, return significantly higher power ratios at the second turbine position, making their results closer to the measured values. Higher power ratios are recorded throughout the farm for models including the Coriolis parameter compared with models without the parameter, suggesting the wakes are being at least partially directed away from downstream ADs. In addition, the application of a non-neutral thermal gradient representing the free atmosphere

Table 1.13 List of four URANS configuration options

Option name	Coriolis parameter	Free atmosphere thermal gradient	Symbol
Option 1	Off	Off	Triangle
Option 2	Off	On	Solid circle
Option 3	On	Off	Square
Option 4	On	On	Diamond

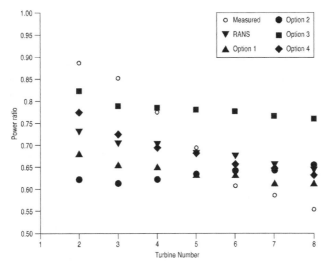

Figure 1.24 Results from the different options for URANS model simulations of case N278 ('+' symbols) using the symbols described in Table 1.13

Measured values from the Nysted wind farm are shown as open circles. For comparison, the results from k-ε RANS simulations are also displayed (inverted triangles).

in the upper domain appears to affect the wake recovery process despite starting hundreds of metres above the turbines, with results from Option 4 displaying a much greater deep-array effect than an otherwise identical Option 3. It is also noted that whilst the inclusion of a thermal gradient in a Coriolis affected simulation appears to strengthen the deep-array effect, its inclusion in a model which excludes the Coriolis parameter appears to weaken it and even aid wake recovery. These results indicate that both the Coriolis parameter and a non-neutral thermal gradient in the free atmosphere should be incorporated in URANS models. Whilst comparisons of results from Options 3 and 4 imply the inclusion of a simulated stable free atmosphere may influence modelled deep-array effects, its height is based on the assumption of neutral stability conditions in the ABL. More fieldwork needs to be conducted to determine its exact height under various conditions and how its fluctuations influence the wake losses in large farms.

Figure 1.25 shows URANS simulations using Option 4 compared with normalised power output for the four offshore wind farms. The results show clear improvements over other models for the Robin Rigg simulations. However, there is little other improvement with URANS showing similar or worse results than the Modified k-ε turbulence model. For the Horns Rev farm, the shape of the graph is similar to that of the Modified k-ε, but with the exception of the free-stream turbine, the URANS predictions are slightly above the power ratios observed. The same can be said of N278 while URANS predicts a slight deep-array effect rather than the gentle recovery seen in N180. The Scroby Sands graph suggests slightly worse power predictions than the Modified k-ε results for wake-effected turbines although each individual machine is modelled well. Across all

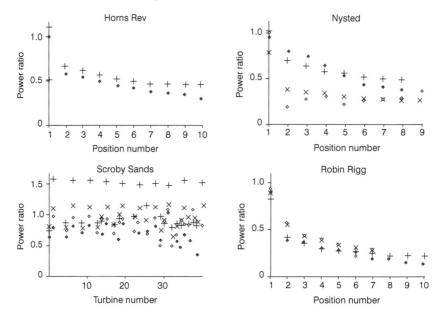

Figure 1.25 Results from the URANS simulations of the four farms.

The symbols are the same as in Figure 1.21.

seven case studies, URANS predicts lower power generation for turbines in the free-stream than any of the other three models with N180, RR204 and RR249 all under-performing the measured data. Combining the decrease in free-stream production with the increase in wake-affected production for URANS suggests the Coriolis force plays a significant role in farm efficiency. Whilst maintaining a constant hub height wind speed for each key turbine, the Coriolis-caused Ekman spiral will alter the vertical wind shear over each AD. While the increase in production deep into N278 and RR204 could be attributed to farm-edge effects with changing wind directions, this explanation is shown not to be the case for HR270 where data from turbines on the farm sides were discarded both in measured data and simulations.

Model comparison

It can be deduced from the previous sections that there are considerable differences between predicted and observed turbine output depending on farm location, layout and model configuration. For example, while the k-ε model performed well with the Horns Rev and Nysted farms, it did less well with the Robin Rigg farm. This may be a due to differences in wind farm layout, turbine size or even the technique used to measure meteorological conditions, variables which are hard to investigate using field data from just the two Danish farms and particularly flow directions 270 degrees and 278 degrees respectively. Similarly, there was variation in accuracy for each model within each farm, specifically Scroby Sands where all four models were more accurate at predicting SS77 than SS90. To help compare the results, Figure 1.26 and Figure 1.27 show the Root Mean Squared (RMS) and the standard deviation (σ) of the prediction errors for each of the seven case studies split by model configuration. Case RR249 provided an unusual comparison as turbines from this direction are staggered and so spaced roughly one diameter further apart downstream than case RR204, with turbines located in the gap between previous turbines in neighbouring rows (see Figure 1.16). This allows the turbine rows to be considered in numerous ways, yet since both measured and predicted generation data were averaged using the same method, simulations produced results directly comparable with reality, with the URANS model the most accurate.

Figures 1.26 and 1.27 show that the SST option is the least accurate overall, as well as showing the most variability within individual farms. There is little separating the other three models for greatest RMS accuracy if only the most popular industry test cases (HR270 and N180) are considered, however, the URANS runs resulted in greatest accuracy in five of the seven case studies with similar RMS values to the most accurate model for the other two cases. URANS also achieved lowest σ values in three cases, only being surpassed by the Modified k-ε. model which achieved the lowest σ values in four cases (where the URANS values for σ were very similar).

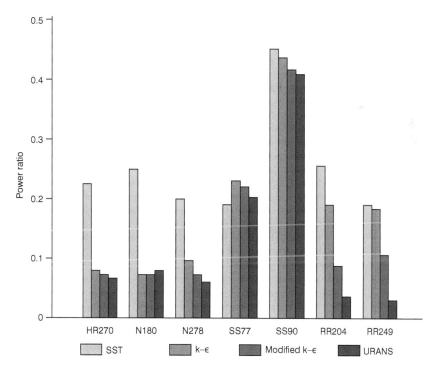

Figure 1.26 Root Mean Squared (RMS) error for each of the seven cases simulated for the four different model configurations

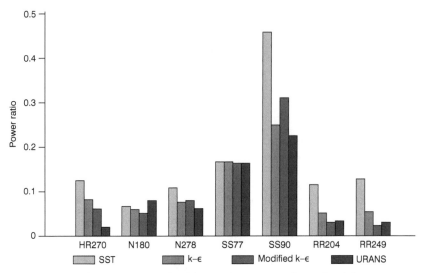

Figure 1.27 Standard Deviation (σ) of the prediction error for each of the seven cases simulated for the four different model configurations

Wind tunnel simulations of offshore farm wake losses

Experimental description

To obtain a better insight into the structure of offshore wind turbine wakes, model 1:300 scale turbines were manufactured and used to make detailed wake measurements using the EnFlo meteorological wind tunnel at the University of Surrey. Detailed information on the experiments can be found in Hancock (2013, 2014a, 2014b), Hancock and Farr (2014).

A diagram of the wind tunnel is shown in Figure 1.28. The tunnel has dimensions in its working section of 1.5 m high, 3.5 m wide and 20 m long and is capable of producing wind speeds of 0.3 m/s to 3 m/s. Velocity measurements are made using a Dantec two-component 40 MHz frequency-shifted laser Doppler anemometry (LDA) system using seeding particles produced from a sugar solution.

The wind tunnel is capable of reproducing environmental flows using Irwin (1981) spires and floor-mounted elements to simulate the desired surface roughness, wind shear and turbulence characteristics following ESDU guidelines (ESDU 2001, 2002). In addition, the flow can be stratified to produce stable, neutral or unstable boundary layers using heating and cooling elements. The issue of scaling from full to model size is discussed in Hancock and Pascheke (2014a).

The model turbines are shown in Figure 1.29 in two different configurations. Each turbine comprises a three-blade rotor with a diameter (D) of 416 mm. The rotor speed is tightly controlled by means of a four-quadrant controller connected to a micro-motor via a gear box and the turbine nacelle is mounted on a solid steel tower 13 mm in diameter. The hub height of each machine is 300 mm. The turbine was developed as an approximation to the SUPERGEN Wind 5 MW

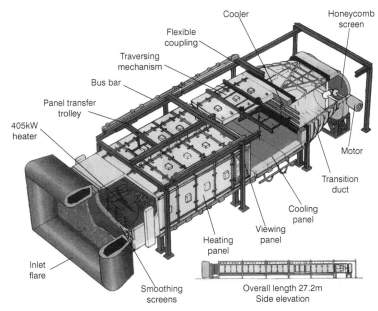

Figure 1.28 A diagram of the EnFlo meteorological wind tunnel at the University of Surrey

exemplar turbine used by the consortium for a number of studies. The turbine blade was made of carbon fibre in the form a twisted thin flat plate, 0.8 mm thick. The rotor was designed to operate optimally at a tip speed ratio of 6.

The wind tunnel was configured to carry out three sets of simulations under stable, neutral and unstable surface layer stability conditions. From previous research (Argyle 2014), it has been shown that for a significant fraction of the time, surface layer stability at offshore sites is non-neutral. This has implications for wake losses due to the effect of differing wind shear, buoyancy and turbulence intensity under different stability conditions.

Figure 1.29 Scale model wind turbines being tested as a single row of four machines (left) and as an array of three by four machines (right)

The parameters corresponding to the different stability conditions simulated in the wind tunnel are given in Table 1.14 showing ABL height (h), surface roughness length (z_0), Obukhov length (L) and overlying inversion strenth ($d\theta/dz$). The wind speed in the tunnel was ~1.5 m/s, corresponding to a full-size turbine wind speed of 10 m/s at hub height. The turbine thrust coefficient for the stable case was 0.48 and for the unstable case 0.42. The neutral simulation was repeated at these two thrust coefficient settings for comparison.

Results

Figure 1.30 shows the upstream profiles of wind and turbulent wind speed conditions for the three different stability conditions simulated. Measurements were repeated three times in each case showing three sets of points for each stability condition. It can be seen that the unstable mean wind speed profile shows a sharp surface layer wind shear initially very close to the ground which quickly levels out. The level of turbulence is greatest in this case. The stable and neutral mean wind speed profiles exhibit similar profiles with a slightly greater shear in the case of the stable situation. The stable simulation shows the lowest level of turbulence.

Table 1.14 Parameters used for the wind tunnel stratified simulations

	Neutral	Stable	Unstable
h (mm)	≈ 1050	≈ 500	≈ 1200
z_0 (mm)	0.10 +/– 0.01	0.11	0.1
h/L	0	0.4	–1.26
$d\theta/dz$ (K/m)	–	≈ 20	≈ 3

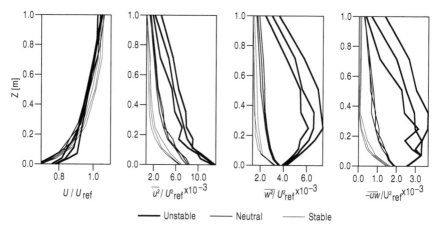

Figure 1.30 Mean wind speed and turbulence profiles used for stable (dark grey), neutral (black) and unstable (light grey) wind tunnel simulations. Values are normalised to the reference wind speed of 1.5 m/s

Figure 1.31 shows vertical mean wind speed and turbulence profiles between 0.5D and 10D downstream of a single turbine, comparing neutral and stable conditions. Measurements of the reference upstream profile (solid lines) are shown for comparison. Comparing the neutral and stable mean wind speed profiles, it can be seen that the momentum deficit is larger in the stable case at each corresponding distance downstream of the turbine. It can also be seen that the wake effect (velocity deficit) takes longer to diminish.This effect was also observed from field measurements by Magnusson (1994). The other major effect of stability is that the vertical growth rate is heavily suppressed by the inversion strength. The upstream turbulence levels are significantly different as was also seen in Figure 1.30. This has consequences for the levels of turbulence in the wake for both the added turbulence and the absolute turbulence which are generally lower in the stable case. The heights at which peaks occur are lower in the stable case, and this is also the case for the height of maximum mean velocity deficit.

Figure 1.32 shows a similar set of profiles to Figure 1.31, this time comparing a neutral and unstable surface boundary layer. In contrast to the neutral and stable cases, it can be seen that the turbulence levels in an unstable (convective) ABL are significantly larger, even though the flow is only slightly unstable (a point that is discussed in Hancock, 2013). The strength of the mean velocity deficit decreases more rapidly in this unstable case compared with neutral. This is as would be expected due to the higher level of turbulence and thus greater mixing in of the faster moving air outside the wake. This is also observed in the field studies of Magnusson (1994). The height of the boundary layer is also increasing more rapidly, as can be seen from the mean velocity and turbulence profiles.

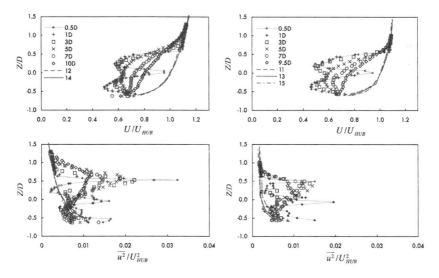

Figure 1.31 A comparison of mean and turbulence vertical profiles at several locations downstream of a single turbine for neutral (left) and stable (right) surface layer stratification conditions

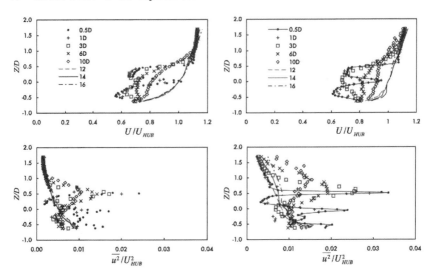

Figure 1.32 A comparison of mean and turbulence vertical profiles at several locations downstream of a single turbine for neutral (left) and unstable (right) surface layer stratification conditions

Wake losses from multiple turbines

The wind tunnel was set up to simulate a neutral surface layer for a row of four turbines 6D apart (see Figure 1.29, left). A second simulation to study the effect of adjacent rows of turbines was configured with two adjacent rows 2.4D either side of the initial row (see Figure 1.29, right). The relatively close lateral spacing of 2.4D was used to ensure the promotion of downwind wake interaction of the three rows, which would not have been possible using a larger spacing in this wind tunnel. In addition, this spacing effectively makes the side walls into symmetry planes (ignoring boundary layers at the walls), thus simulating an infinite array in the lateral direction. Figure 1.33 compares the wake profiles at hub height downstream of the first central row turbines for both the single-row and three-row configurations. All wind speeds are normalised to the wind tunnel reference wind speed. Between 7D and 13D, the effect of the adjacent turbines can be seen in a slight increase in the velocity at the edge of the wake due to the increased blockage. The velocity minima at these two distances downstream are also slightly higher due to the higher velocity external to the wake as a result of the blockage. At 4D downstream of the first turbine, there is a region at the edge of the wake in which the wind speed is almost constant, but by 7D this is no longer the case, implying that the adjacent wakes are impinging on the wake of the second central turbine. The interaction is strong enough for the maxima at 16D to reduce to below those at 10D, and the maxima at 22D to reduce to below those at 16D. The minima at these latter positions downstream are still slightly higher. It can also be seen that the wakes are slightly narrower as a result of the blockage of the adjacent wakes.

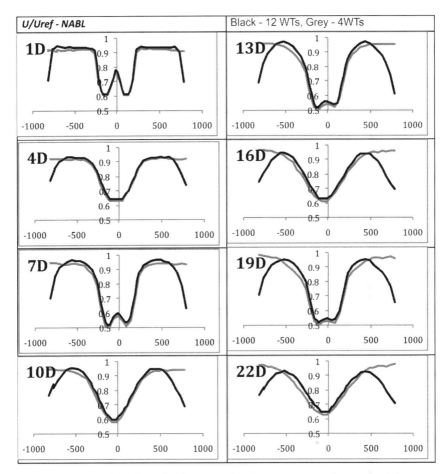

Figure 1.33 Normalised lateral velocity profile (x-axis is in mm) at several distances downstream of a single row of four turbines (grey) and downstream of an array of three by four turbines (black) in the case of a neutral surface layer, Farr (2014)

References

Argyle P and Watson SJ, 2012. A study of the surface layer atmospheric stability at two UK offshore sites. *Scientific Proceedings of European Wind Energy Association Annual Event*, Copenhagen, 16–19 April 2012, pp. 110–113.

Argyle P and Watson SJ, 2014. Assessing the dependence of surface layer atmospheric stability on measurement height at offshore locations. *Journal of Wind Engineering and Industrial Aerodynamics* 181: 88–99.

Aubrun S, Loyer S, Hancock P and Hayden P, 2013. Wind turbine wake properties: comparison between a non-rotating simplified wind turbine model and a rotating model. *Journal of Wind Engineering and Industrial Aerodynamics* 120: 1–8.

Brower M, Vidal J and Beaucage P, 2013. Evaluation of four numerical wind flow models. *Proc. EWEA Technology Workshop: Resource Assessment*, Dublin, 2013.

Carney G, Housley P, Montavon C and Jones I, 2012. Practical usage of CFD for Wake effects prediction within offshore wind farms. *Proc. European Wind Energy Association Conference and Exhibition*, Copenhagen, 2012.

Crasto G and Castellani F, 2013. Wakes calculation in a offshore wind farm. *Wind Engineering* 37: 269–280.

Dumais RE, Passner JE, Flanigan R, Sauter B and Kirby S, 2009. High resolution WRF-ARW studies at the U.S. Army Research Laboratory for use in short-range forecast applications. P2.4. 23rd Conference on Weather Analysis and Forecasting/19th Conference on Numerical Weather Prediction. Omaha, NE, June 1–5, 2009.

ESDU 2001. Report 85020 Engineering Sciences Data Unit, London, UK.

ESDU 2002. Report 82026 Engineering Sciences Data Unit, London, UK.

Farr TD, 2014. The effects of atmosphere and wake turbulence on wind turbines and wind turbine wakes. PhD thesis, University of Surrey.

Flygare ME, Austin JA and Buckwalter RM, 1985. Maximum likelihood estimation for the 2-parameter Weibull distribution based in interval-data. *IEEE Trans. on Reliability* R–34: 57–59.

Gibbs JA, Fedorovich E and van Eijk AMJ, 2011. Evaluating weather research and forecasting (WRF) model predictions of turbulent flow parameters in a dry convective boundary layer. *J. Appl. Meteor. Climatol.* 50: 2429–2444.

Hancock PE, Zhang S and Hayden P, 2013. A wind tunnel artificially-thickened weakly-unstable atmospheric boundary layer. *Boundary-Layer Meteorol.* 149: 355–380.

Hancock PE and Farr TD, 2014a. Wind tunnel simulation of wind turbine arrays in neutral and non-neutral winds. *Journal of Physics*: conf series S24.

Hancock PE and Pascheke F, 2014b. Wind tunnel simulation of the wake of a large wind turbine in a stable boundary layer: Part 2 the wake flow. *Boundary-Layer Meteorol.* 151: 3–21.

Hancock PE and Pascheke F, 2014. Wind tunnel simulation of the wake of a large wind turbine in a stable boundary layer: Part 1, the boundary layer simulation: Part 2 the wake flow. *Boundary-Layer Meteorol.* 151: 23–37.

Hancock PE and Zhang S, 2015. A wind-tunnel simulation of the wake of a large wind turbine in a weakly unstable boundary layer. *Boundary-Layer Meteorology* 156(3): 395–413.

Hong SY, Noh Y and Dudhia J, 2006. A new vertical diffusion package with explicit treatment of entrainment processes, *Mon. Weather Rev.* 134: 2318–2341.

Irwin HPAH, 1981. The design of spires for wind simulation. *Journal of Wind Engineering and Industrial Aerodynamics* 7: 361–366.

Janjic ZI, 2001. Non-singular Implementation of the Mellor–Yamada level 2.5 scheme in the NCEP meso model. NCEP Office Note No. 437, p. 61.

Janjic ZI, 2003. A non-hydrostatic model based on a new approach. *Met. Atmos. Phy.* 82: 271–285.

Jiménez PA and Dudhia J, 2012. Improving the representation of resolved and unresolved topographic effects on surface wind in the WRF model. *J. Appl. Meteor. Climatol.* 51: 300–316.

Litta AJ and Mohanty UC, 2008. Simulation of a severe thunderstorm event during the field experiment of STORM programme 2006, using WRF-NMM model. *Current Sci.* 95: 204–214.

Magnusson M and Smedman A-S, 1994. Influence of atmospheric stability on wind turbine wakes. *Wind Engineering* 18: 139–152.

Mass C and Ovens D, 2011. Fixing WRF's high speed wind bias: A new subgrid scale drag parameterization and the role of detailed verification. Preprints, 24th Conf. on

Weather and Forecasting/20th Conf. on Numerical Weather Prediction, Seattle, WA, Amer. Meteor. Soc., 9B.6. Available online at http://ams.confex.com/ams/91Annual/webprogram/Paper180011.html [Accessed 13/09/2013].

Menter F, Kuntz M and Langtry R, 2003. Ten years of industrial experience with the SST Turbulence Model. *Proc of the 4th International Symposium on Turbulence, Heat and Mass Transfer*, Antalya, 2003.

Montavon C, 1998. Simulation of atmospheric flows over complex terrain for wind power potential assessment, PhD thesis No. 1855, Lausanne, EPFL, Switzerland.

Montavon C, Hui S, Graham J, Malins D, Housley P, Dahl E, de Villiers P and Gribben B, 2011. Offshore wind accelerator: wake modelling using CFD. *Proc. European Wind Energy Association Conference and Exhibition*, Brussels, 2011.

Nunalee CG and Basu S, 2014. Mesoscale modeling of coastal low-level jets: implications for offshore wind resource estimation. *Wind Energy* 17: 1199–1216.

Peña A, Hahmann AN, Hasager CB, Bingöl F, Karagali I, Badger J, Badger M, and Clausen NE, 2011. *South Baltic Wind Atlas*. South Baltic Offshore Wind Energy Regions Project. Risø, DTU.

Rados K, Prospathopoulos J, Stefanatos N, Politis E, Chaviaropoulos P and Zervos A, 2009. CFD modeling issues of wind turbine wakes under stable atmospheric conditions. *Proc. European Wind Energy Assosiation Conference and Exhibition*, Marseilles, 2009.

Réthoré P, Sørensen N and Zahle F, 2010. Validation of an actuator disc model. *Proc. European Wind Energy Assosiation Conference and Exhibition*, Warsaw, 2010.

Saha S et al., 2010. The NCEP Climate Forecast system reanalysis. *Bull. Amer. Meteor. Soc.* 91: 1015–1057.

Seshadhri S, Montavon C and Jones I, 2013. Coriolis effects in the simulation of large array losses: preliminary results on 'wall effect' and wake asymmetry, in *Offshore '13*, Frankfurt, 2013.

Skamarock WC, Klemp JB, Dudhia J, Gill DO, Barker DM, Duda MG, Huang X-Y, Wang W and Powers JG, 2008. A description of the advanced research WRF Version 3. NCAR/TN–475=STR, NCAR Technical Note, Mesoscale and Microscale Meteorology Division, National Center of Atmospheric Research, June 2008, 113pp.

Tastula E-M, Vihma T and Andreas EL, 2012. Evaluation of polar WRF from modeling of the atmospheric boundary layer over Antarctic sea ice in autumn and winter. *Mon. Wea. Rev.* 140: 3919–3935.

Uppala SM et al., 2005. ERA–40 Re-analysis. *Quarterly Journal of the Royal Meteorological Society* 131: 2961–3012.

Zhao P, Wang J, Xia J, Dai Y, Sheng Y and Yue J, 2012. Performance evaluation and accuracy enhancement of a day-ahead wind power forecasting system in China. *Renewable Energy* 43: 234–241.

Zilitinkevich S and Mironov D, 1996. A multi-limit formulation for the equilibrium depth of a stably stratified boundary layer. *Boundary-Layer Meteorology* 81: 325–351.

2 Environmental interaction

Laith Danoon, Anthony K. Brown,
Vidyadhar Peesapati and Ian Cotton

Introduction

Large onshore turbines and offshore wind farm technology are expected to make a significant contribution to meeting the UK Government target of 10 per cent of the country's electricity needs through renewable sources (the *Renewable's Obligation*). However, a significant proportion of new and existing wind farm planning applications submitted to regional agencies for consideration have objections raised to them on the basis of the potential for radar interference

The interference of wind farms with aviation and marine radar is considered a significant concern to the regulating authorities. In the UK in particular, and Europe in general, strict rules concerning radar interference have caused planning applications to be rejected, modified or significantly delayed (Poupart, 2003).

The majority of planning applications for "Round 1" wind farm sites were submitted in the late 1990s to 2000. Many of these were for relatively small numbers of what are now considered to be smaller turbines, predominantly onshore or coastal based due to the additional generating costs of remote offshore development [Pinto 2006]. Despite this, a number of sites either remain at the planning stage or have since been abandoned following planning objections raised on the basis of microwave radar or communications interference. Similarly, some of the more recent "Round 2" and "Round 3" applications, which are usually for larger wind farms with larger turbines, typically located offshore, have objections lodged to their planning applications due to radar interference concerns.

With increasing turbine sizes, the maximum tip heights and tip speeds are making the turbines more visible to radars. Thus it is anticipated that the problem of objections on the basis of radar interference will become increasingly acute.

This chapter discusses some of the underlying issues and causes relating to the interaction between wind farms and radar systems. The significance of the radar signature of rotating and static turbine components is also highlighted, stating its role in the potential interference with particular types of radars. Finally, the advantages and challenges of applying radar signature reduction techniques to wind turbines are discussed with particular emphasis on their effect on the lightning protection system.

Impact of wind farms on radar systems

Surveillance radars operate mainly between 2.8 – 9.4 GHz. Radars operating at 2.8 – 3.2 GHz or S-Band are mainly for Air Traffic Control (ATC) and Air Defence (AD). These radar systems are high in complexity and they use advanced signal processing algorithms to distinguish between real targets and clutter. One of the techniques used is Doppler processing and filtering to distinguish between moving objects from static and unwanted targets, i.e. clutter.

Radar systems operating at 9.2 – 9.4 GHz, or X-Band, include marine navigational radars and Vessel Tracking Services (VTS). These systems are lower in system complexity and rarely use Doppler processing. Therefore, the main concern for such radars is the large radar-cross section (RCS) that might degrade the performance.

Other radar systems do operate outside these frequency bands such as the en-route radar operating at L-Band and weather radars operating at C-Band. However, in many cases the impact of wind farms can be looked at from two main points of view, Doppler radars and non-Doppler radars.

Non-Doppler radar systems

Wind turbines are complex in geometry and extend over hundreds or thousands of wavelengths at radar frequencies. Their materials, along with the size and overall geometry of the tower, blades and nacelle, make turbines a very good reflector of radar signals. Their RCS, which is a measure of the energy reflected back to the radar, is often comparable to that of a large vehicle carrier or an oil-tanker. The large size and high RCS is the main cause of interference with non-Doppler radars which can cause the following problems.

Receiver limiting

Receiver limiting occurs due to the high radar returns from the wind turbines. The received power can be large enough to cause the dynamic range of the receiver to saturate. This will cause small targets to be lost.

Target smudging and sidelobe detections

Due to the very large RCS of wind turbines, detection can occur in the lower part of the radiation pattern main beam (causing an arc on the display) or in the sidelobes showing split targets and/or the appearance of false targets that are of the same range but at a different azimuth angles (Howard 2004, Marico 2007). In the case of marine navigational radar onboard vessels, Figure 2.1 shows this effect in measurements made near the North Hoyle offshore wind farm.

The ship's structure can add to the severity of these effects. It is noted that when the ship's mast is between the radar antenna and the wind farm, returns from the wind turbines are extended in azimuth and more sidelobe detections

Figure 2.1 Measured data showing sidelobe detections near North Hoyle wind farm

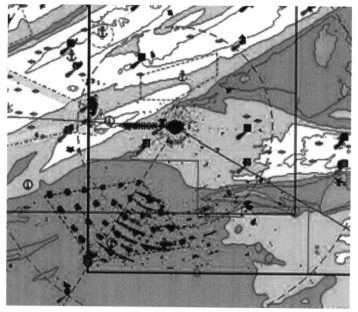

Figure 2.2 Sidelobe detections near Kentish Flats wind farm (Marico 2007)

occur. This is due to the distortions to the antenna radiation pattern caused by structures that are very close to the antenna. An example of this phenomenon can be seen in the measured data shown in Figure 2.2.

The measurement shows that the returns from turbines are severely distorted. The distortion is causing the returns from individual turbines to merge into each other. This results in over-crowding of the radar screen, which can confuse the radar operator and make it difficult to track targets close to the large object.

Shadowing

Offshore wind farms with large number of turbines tend to induce large shadow areas behind the farm area. The severity of this effect depends on several factors, including the geometry of the tower, the wind farm layout and the radar distance from the wind farm.

When turbines are placed within the line of sight (LOS) of radar systems, radar shadowing will occur behind the structure. The extent and length of the shadow region depends on the size of the turbine, the distance to the radar antenna, the height of the radar and the height of the target of interest. The severity of the shadow will also depend on the distance of the target from the turbine. Radar diffraction around the turbine will result in a reduced effect of the shadow as the range between the shadowed target and the turbine is increased. This is an

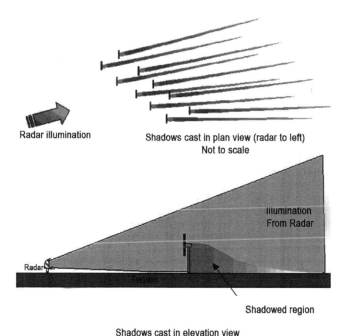

Figure 2.3 Illustration of radar shadowing with diffraction effects (Butler 2003)

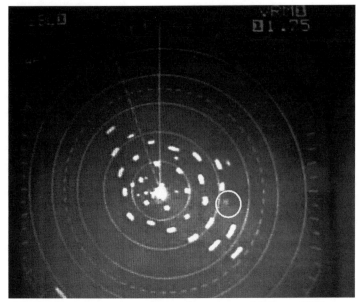

Figure 2.4 Radar shadow causing a turbine to be undetected (Marico 2007)

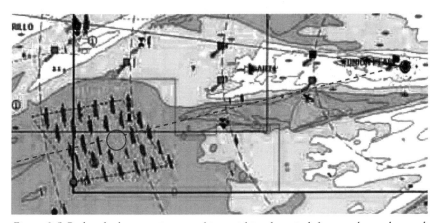

Figure 2.5 Radar shadow causing a turbine within the wind farm to be undetected (Marico 2007)

important characteristic of the radar shadow and is illustrated in Figure 2.3. It has been reported that a target 1 km behind the turbine will experience a 6dB reduction in the returned power while targets that are significantly further away suffer only a 2dB reduction in the received radar echo (Butler 2003).

Figures 2.4 and 2.5 are measured data showing a turbine being undetected due to the shadow induced by other turbines.

The shadows caused by wind farms may cause the disappearance of moving targets when they enter the shadow region. For marine radars mounted on moving vessels, the effect of the shadowing is variable depending on the position of the vessel. These effects are often transient provided that the target is at a sufficient distance from the wind farm (Marico 2007).

Multiple reflections

When the marine radar is operating close to a wind farm the radar signals can be reflected by nearby targets or objects, the ship structure and between the turbines within the farm. These multiple reflections of the radar signal can cause the appearance of ghost targets on the radar screen. This may cause the tracker to lock on these spurious targets and to initiate new tracks and targets.

The ship structure may cause the radar signal to be reflected to the wind farm and back to the radar. It has been shown that positioning the radar antenna near large reflecting surfaces may cause the appearance of large mirror images of the wind farm (Marico 2007). These types of multiple reflection effects may cause severe cluttering of the radar display. When the radar is moving, the position of the ghost targets will vary, potentially masking real targets.

Multiple reflections may also occur when another large object is near the radar and the wind farm such as a large offshore structure, terrain features or large vessels in the vicinity. These large nearby objects and targets may reflect the radar signal towards the wind farm and back to the receiver, causing the appearance of a line of ghost targets in the direction of the reflecting object.

It has been observed that when the radar is operating within close proximity of a wind farm, spurious targets appear on the radar display. These ghost targets appear due to multiple bounces of the radar signal within the wind farm. Figures 2.6 and 2.7 show measurements made near the North Hoyle wind farm. The appearance of a line of ghost targets can be seen in the direction of the nearby turbines.

The appearance of ghost targets due to the multiple reflection of the radar signals can cause cluttering of the radar display and mask real targets. These effects can be a threat to ships operating within close proximity of the wind farm in poor visibility. The cluttering of the radar display of regulating and security authorities may also be used by small vessels to avoid detection.

A combination of the above effects may degrade the tracker performance of safety critical radar systems offshore such as the Radar Early Warning Systems (REWS) installed on offshore oil and gas installations. Such systems are used to detect and track vessels in the vicinity and alert the operator for vessels that are on a collision course with the platform. These systems must maintain strict detection and tracking standards in order to ensure the safety of the personnel on the platform and for asset management and safety.

Another radar system that does not usually use Doppler in its processing is the Secondary Surveillance Radar (SSR). The SSR is a cooperative system that relies on transponders onboard aircrafts to give information back to the system. The ground-based end of the system sends a set of pulses which interrogate the surrounding airspace. Aircraft with suitable transponders will automatically respond to such interrogations and will relay back information from the aircraft including its altitude and identification.

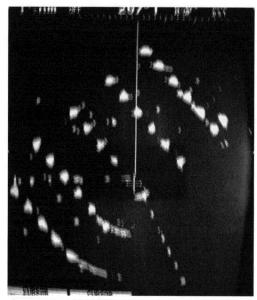

Figure 2.6 Linear ghost target detections caused by multiple reflections within the wind farm (Rashid 2007)

Figure 2.7 Detection of ghost targets due to multiple reflections from turbines when close to the wind farm (Rashid 2007)

Although the SSR is a robust and well proven system, like any radar, it can be affected by poor siting. Obviously, blockage by local structures will reduce coverage. Multipath reflections off the ground may also be an issue. However, there is one further issue to do with near field blockage of SSR. The presence of a nearby structure can cause the null location to be misplaced with respect to the sum pattern when near an obstacle. This therefore results in an angular error in the reported position of targets as detected.

SSR coverage may be up to 250 nm and this potential error may represent a large cross-range deviation at the extremes of range. To this author's knowledge, this effect has not been reported in general but may be important in extreme cases such as a nearby hanger being close to an SSR installation. Nonetheless the effect of nearby obstacles on an SSR cannot be considered negligible.

Measurements and reports show that wind turbines do indeed interfere with non-Doppler radars used in offshore environments. Unlike air traffic control radars, non-Doppler radars are often low complexity/cost. The practicality of introducing advanced signal processing into these radars to reduce wind farm impact is considered unlikely.

Radar systems with Doppler processing

The civilian Primary surveillance (and many AD) radars (PSR) use short pulses of radio frequency energy. By timing the interval between when a pulse is transmitted and received the radar determines the range of any reflecting object within the operational range of the radar. No attempt is made to determine elevation position (altitude) in standard ATC primary radars. PSR has the major benefit that it will detect any target within its operational parameters including targets that may be non-cooperative. In many countries (though not the UK) it is common to have the SSR antenna co-mounted on the primary radar. Otherwise the SSR systems may be physically located away from the PSR. In either configuration the primary radar system works in conjunction with data from an SSR system. This produces an accurate and overall representation of the activity within a defined airspace.

AD and ATC radars employ Doppler processing to distinguish between moving targets of interest and static clutter. In addition to the issues that are associated with non-Doppler radars, such as shadowing, receiver limiting and multiple reflections, Doppler radars may suffer due to the nature of the turbine blades, their wind-dependent orientation and their rotation. The interference due to the rotating blades and the large reflection of the radar signal has been well reported and explained in the literature. The main cause of concern is the large RCS of the wind turbines and the Doppler frequency shift caused by the rotating blades, which may appear to the radar as a moving aircraft. These unpredictable detections of the blade tips over a wind farm area may cause false targets to appear or real targets to be lost.

ATC and AD radars also employ tracking as part of their data processing. As a consequence of the variable Doppler returns, wind turbines may also generate false tracks or cause successfully established tracks to be lost or altered as targets approach wind farms.

Additionally, AD and ATC radars generate clutter maps, where each clutter cell is composed of returns from a number of range bins and the highest return is used to set the clutter level for a particular cell containing multiple range bins. In the case of small-target detection in the vicinity of wind farms, this has the unfortunate effect of raising the level over the whole of the clutter cell consisting of a number of range bins, in which one wind turbine might be located.

As ATC radars and some AD radars do not account for the target altitude, many existing radars are not able to discriminate targets in, near or above wind farms, as shown in Figure 2.8. However, a number of techniques can be adopted to improve this situation. One obvious potential solution would be to tilt the antenna in elevation so that the main beam misses the wind farm, as shown in Figure 2.9. However, this is likely to result in a concomitant reduction in sensitivity to targets at low altitudes which may not be acceptable in the case of AD radars. In addition, introducing antenna tilt risks bringing beam sidelobes to an elevation where they illuminate the wind farm. The tilt angle needed could be minimised (and so low altitude detection capability preserved) if the beam were very narrow. However, this is impractical because a beam with a sharp roll off in elevation would require a very large aperture to generate it.

Wind turbine radar scattering characteristics

As has been shown so far, the main cause of concern is the large RCS of the wind turbines and the Doppler effect from the rotating blades. Due to their large size and construction material, wind turbines are excellent reflectors of radar energy. There have been a number of studies to model and measure the turbines' RCS and to use such data to predict the impact of a specific wind farm on a

Figure 2.8 Overhead obscuration over the wind farm area for 2D ATC and AD radars

Figure 2.9 Applying antenna tilt to miss the wind farm area

radar system. RCS analysis and scenario is often considered to be the most cost-effective method to assess the potential impact of wind farms on radar.

The models developed with the Supergen programme were used to model the total RCS of a generic wind turbine shown in Figure 2.10. A tilt angle of 5° is applied to the blades to provide sufficient clearance distance between the rotating blades and the tower. As a common measure with RCS calculations, the radar is placed within the farfield of the turbine and the turbine is assumed to be rotating at 15 rpm. Furthermore, the turbine structure is assumed to be a perfect electrical conductor (PEC) shell and the RCS results were modelled at 3 GHz which is representative of what ATC and AD radars operate at.

The highest RCS profile is expected to occur when the turbine is facing the radar. This is due to the illumination of the large and relatively flat pressure-sides

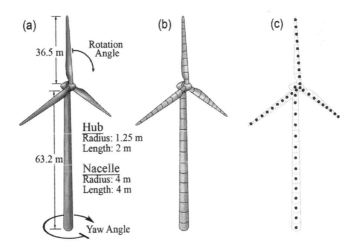

Figure 2.10 (a) CAD geometry of a generic 2 MW turbine used in the RCS modelling, (b) segmented turbine for RCS modelling, (c) The RCS point located at the centre of each segment

of the blades. Figure 2.11 shows the total turbine RCS generated over one full rotation. Figure 2.11 also shows the RCS of each component individually.

The results show that for this particular turbine geometry, the contribution from the cylindrically tapered tower is 20 dBm², which will be constant at all yaw angles. The blades are the dominant contributor to the total RCS at this illumination angle reaching peak values of 44 dBm². At this yaw angle, the Doppler is low as the blades are not moving significantly in the direction of the radar. Figure 2.12 shows the Doppler signature generated by the blades at 0° yaw. Only the contribution from the blades is modelled as the tower and the nacelle are assumed to be static and are not moving or vibrating as the blades rotate. Although the RCS levels are high, reaching up to 40 dBm², the Doppler shift is relatively low and is confined between ±120 Hz. This is caused by the tilt applied to the blades to avoid collision with the tower, which generates a slight movement towards/away from the radar as the blades rotate.

The illumination of the side of the turbine is considered to be the worst-case scenario for Doppler-based radar systems. Therefore the ReMeRA model was used to predict the RCS of the turbine at 90° yaw. The results of the static RCS profile are shown in Figure 2.13 while the Doppler signature from the blade is shown in Figure 2.14.

The static RCS of the turbine is dominated by returns from the nacelle at most rotation angles. The tower is also contributing significantly to the total RCS while the blades are only flashing when the leading-edge of a blade is illuminated. The Doppler signature plot in Figure 2.14 shows the maximum Doppler shift caused is ±1100 Hz. The illumination of the leading-edge generates high return with a high Doppler shift. When the trailing edge is illuminated, the high returns are generated by the cylindrical root section of the blade which does not contribute to the high Doppler returns.

It is often assumed that the radar is positioned in the farfield and on the same level as the turbine. This assumption is not always valid when considering the

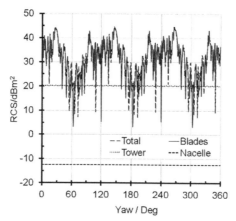

Figure 2.11 Turbine RCS over a complete rotation at yaw = 0°

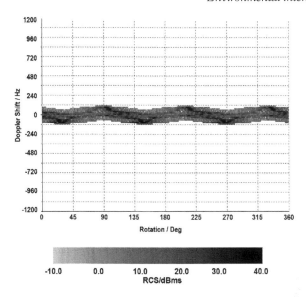

Figure 2.12 Doppler signature of the turbine blades at 0° yaw

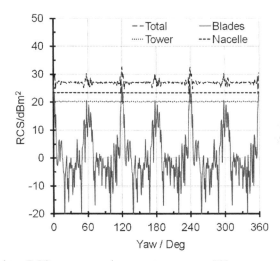

Figure 2.13 Turbine RCS over a complete rotation at yaw ±90°

effects of wind farms on radars. Onshore wind farms may be positioned on top of hills and high ground which may cause a slight angle to the radar antenna. Also, when modelling offshore wind farms, the affected radars may be positioned on high cliffs or on top of towers. Additionally, marine radars may operate within close range of the wind farm, causing an angle of incidence in the elevation plane.

Figure 2.15 show the RCS modelling results of a turbine by varying the elevation angle while keeping the yaw angle at 0°. It can be noted that the

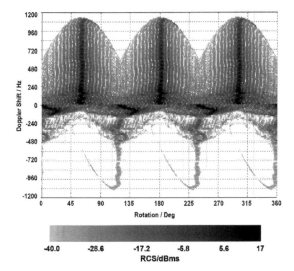

Figure 2.14 Doppler signature of the turbine blades at ±90° yaw

contribution from the tower is only dominant at the specular reflection angle near the horizontal. The rest of the RCS profile is a combination of the scattering from the nacelle and the blades.

When the turbine was modelled at 90° yaw and the elevation angle was varied, the results in Figure 2.16 show that at this particular yaw angle, the RCS of the turbine is dominated by the returns from the nacelle at most elevation angles. The results signify the importance of a radar-friendly nacelle design in order to minimise the static RCS of the turbine at such angles, as shown in the next section of this

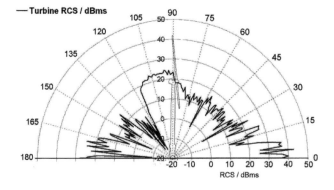

Figure 2.15 Total turbine RCS pattern as a function of elevation incidence angle at 9 GHz (face-on illumination)

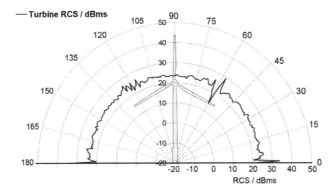

Figure 2.16 Total turbine RCS pattern as a function of elevation incidence angle at 9 GHz (side illumination)

chapter. It also shows the need to accurately represent the geometrical definition of the nacelle when modelling the total RCS of the turbine.

When modelling the RCS of wind turbines, the results are often presented as a farfield set of values. To obtain the farfield results shown previously, the radar source point was assumed to be 500 km away from the turbine structure. It has been noted however that marine navigational radars and some Vessel Tracking Systems (VTS) may operate within a few kilometres of offshore wind farms. Thus, the use of farfield RCS to model the impact of wind farms on radars operating near the wind farm may give inaccurate results. Modelling of nearfield RCS is often needed. The variation in the RCS of the Exemplar 2 MW and the Exemplar 5 MW turbines with respect to radar range are shown in Figure 2.17.

It is clear that it may not be of acceptable accuracy to assume that the scattering at the nearfield is equivalent to that at the farfield. Furthermore, care must be taken when representing the turbine as a point scatterer located at the tip height or at the geometrical centre of the turbine. This might be acceptable at long ranges nearing the farfield distance, but may not be the case when modelling scenarios where the wind turbines are close to the radar or in an environment where partial shadowing due to terrain may occur.

Wind turbine RCS reduction

If the scattered signals from the turbines can be reduced significantly, their effect on radar can be reduced to negligible levels. To reduce the scattering from wind turbines, the use of radar stealth has been considered as a possible mitigation solution. The term "stealth" does not completely refer to the application of Radar Absorbing Materials (RAM). In fact, it largely depends on shaping the target to reduce the reflections of the radar signal.

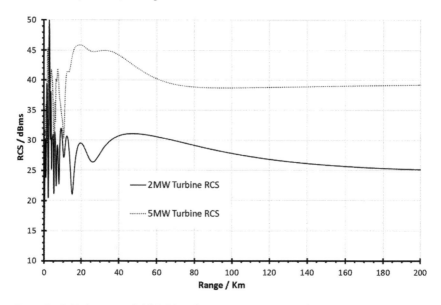

Figure 2.17 Turbine nearfield RCS with respect to range to radar

RCS reduction by shaping

Significant RCS reduction can be achieved through simple alterations to the shape of the object. In the case of wind turbines, reduction of the RCS of the tower and the nacelle is best achieved by shaping the flat reflective surfaces.

In many turbine designs, the tower is considered the largest scatterer of radar signals. This occurs when the tower has a large parallel-sided cylindrical section. For offshore turbines, the large metallic transition piece contributes significantly to the total RCS of the tower. Upright cylinders without any tapering tend to reflect the signal straight back to the radar. Hence, by adding a slight taper to the tower, the radar energy will be directed away from the radar. This is illustrated in Figure 2.18.

Although the nacelle is relatively small when compared with the rest of the turbine, it may contribute significantly to the total RCS as shown in the previous section. The flat and upright sides of the nacelle will act as a good reflecting surface for radar signals when illuminated at broadside. By ensuring that the nacelle sides have a curved or tilted surface, the radar signal would be scattered away from the radar. This will further reduce the RCS of the total turbine at yaw angles of 90° and 270° where ATC radars are most affected. An illustration of the nacelle design is shown in Figure 2.19.

RCS reductions through the application of RAM

The geometry and the airfoil profile of the blade are carefully designed to maximise the wind turbine efficiency. Although RCS reduction through shaping

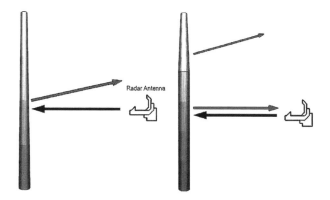

Figure 2.18 The effect of parallel tower and tapered tower on radar reflections

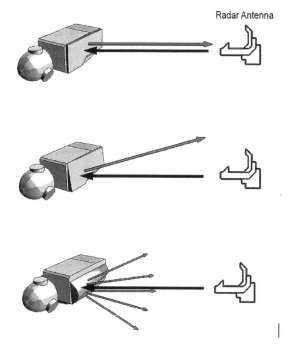

Figure 2.19 Nacelle design and radar reflections

is very effective for other parts of the turbine, changing the shape of the blade is not possible. Therefore, in order to reduce the RCS of the blades, the application of RAM must be considered.

The application of RAM to reduce the interaction with radar has been around for many years. Many solutions are capable of withstanding extreme weather conditions and have a long lifespan. Depending on the application, RAM can be designed to give very good absorption at narrow frequencies, or a reduced performance but over a wide band of frequencies.

To avoid any alterations to the blade profile, RAM solutions must be incorporated into the existing blade design. This requires a thin and lightweight RAM solution. Most compact RAM solutions often need to be placed over a thin metallic reflective layer such as the widely used Salisbury screen absorber. Introducing a metallic layer into the blade construction will make the blade more susceptible to lightning strikes and reduce the efficiency of the integrated lightning protection system. However, it has been commonly suggested that treating certain parts of the blade with RAM while leaving the rest of the blade surface untreated may minimise the added weight penalties and retain the efficiency of the lightning protection system.

Wind turbine and lightning protection

Lightning is an atmospheric discharge of current. The highest recorded value of lightning current is around 250 kA (Berger 1967, 1975). However, this value is very rarely seen, the median (for a downward negative stroke) being about 30 kA with the median values of charge transfer and specific energy being 5.2 C and 55 kJ/Ω respectively (Berger 1967, 1975). The visible part of the lightning strike process, whether lightning strikes the ground or not, is termed as a "lightning flash". The individual components of this lightning flash are defined as strokes. Lightning can be classified into two main types, downward and upward initiated. These are also known as cloud-to-ground and ground-to-cloud lightning, respectively. These two forms of lightning can be further subdivided into positive and negative polarity, the polarity being that of the charge transferred from the cloud to the ground.

Downward-initiated lightning

Downward-initiated lightning starts from the cloud with a stepped leader moving towards the earth. The end of the leader, the leader tip, is in excess of 10 MV with respect to the earth. As the tip gets near to the earth, it raises the electric field strength at the surface of the earth. Where this field is elevated significantly, typically around sharp and/or tall objects, answering leaders are emitted and travel towards the downward-propagating leader. When an answering leader and stepped leader meet, this completes the channel or path from the cloud to earth, thus allowing the charge in the cloud to travel through the ionised channel. This is the first return stroke, and has a peak value of up to a few hundred kiloamps and a typical duration of a few hundred microseconds. After a certain time interval, further strokes may follow the already ionised path and are known as subsequent return strokes, as shown in Figure 2.20. On average, a negative downward lighting flash may contain 2 to 3 subsequent return strokes. Positive downward flashes (only 10 per cent of those observed worldwide) are higher in magnitude but typically contain no subsequent strokes.

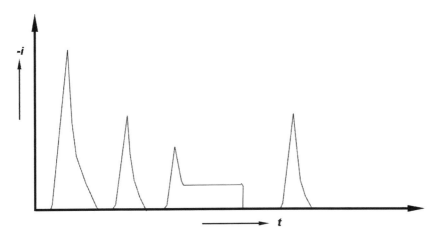

Figure 2.20 Profile of downward-initiated lightning

Upward-initiated lightning

The presence of tall structures and objects brings rise to another form of lightning, which is upward initiated. Tall structures enhance the electric field produced by the thundercloud and this can give rise to upward-propagating leaders that move towards the cloud and which can then develop into a lightning flash. This phenomenon is particularly common where the cloud height is low (often during winter months in coastal areas or in mountainous regions). The profile of an upward discharge is different to that of a downward-initiated discharge, as shown in Figure 2.21. An upward-initiated discharge starts with a continuing current phase on which may be superimposed short duration high magnitude current pulses. Though the current values are quite low at around 10 kA as compared with downward lightning, the charge transfer associated with the continuing current phase can be quite high. The initial continuing current in upward-initiated lightning may be followed by a number of return strokes that are similar to those observed in a negative downward lightning flash.

Lightning current and magnitudes

Lightning parameters are now covered in the new IEC 61400 – 24 'Lightning Protection of Wind Turbines' standard. The parameters for downward-initiated and upward-initiated lightning are given in Tables 2.1 and 2.2.

The lightning effects on wind turbines can be mainly classified into two types:

- direct effects and
- indirect effects.

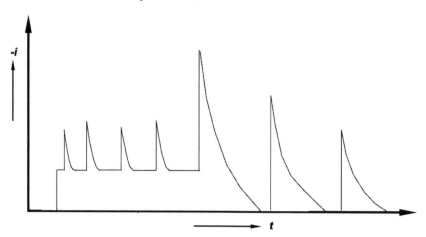

Figure 2.21 Profile of upward-initiated lightning

Direct effects are seen as burns, erosion and damage to the structure due to arc attachment. Indirect effects are those which are due to electromagnetic fields that accompany lightning strikes. The lightning parameters that cause direct and indirect effects are peak currents, specific energy, rate or current rise and charge transfer.

Table 2.3 shows how each of these parameters affects different parts of the wind turbine (Cotton 1998).

From Table 2.3 it can be seen that lightning strikes are capable of causing damage to all the main components of a wind turbine. To minimise and limit the damage caused due to lightning attachment, most wind turbines include lightning protection systems to components that are at risk.

Table 2.1 Downward-initiated lightning parameters

Parameter	Values			Type of stroke
	95%	50%	5%	
I (kA)	4(98%)	20(80%)	90	*First negative short
	4.9	11.8	28.6	*Subsequent negative short
	4.6	35	250	First positive short (single)
W/R (kJ/Ω)	6	55	550	First negative short
	0.55	6	52	Subsequent negative short
di/dt_{max} (kA/µs)	25	650	15 000	First positive short
	9.1	24.3	65	*First negative short
	9.9	39.9	161.5	*Subsequent negative short

Table 2.2 Upward-initiated lightning parameters

Parameter	Units	Maximum value
Total charge transfer	C	300
Total duration	s	0.5–1.0
Peak current	kA	20
Average rate of rise superimposed impulse currents	kA/ μs	20
Number of superimposed impulse currents		50

Table 2.3 Lightning parameters and endangered components of wind turbines

Lightning parameter	Effect	Endangered components
Peak current (kA)	Direct	Blades, lightning protection system, and equipment in nacelle
Specific energy (J/Ω)	Direct	Blades, lightning protection system, and equipment in nacelle
Rate of current rise (A/s)	Indirect	Control and electrical system of wind turbine
Charge transfer (C)	Direct	Bearings and blades

Lightning and wind turbines

The high peak currents carried by lightning strikes are a source of significant energy. If the wind turbine lightning protection system does not divert this lightning current safely to ground through a low impedance path, significant damage can result. If a component of a wind turbine is damaged there are two things that need to be considered: the resulting repair costs and the associated loss in production caused due to downtime. Given the move of wind turbines to offshore locations and the need to mobilise special floating cranes, these costs and repair times have increased in recent years. Because of the risk of damage and possible downtime due to lightning strikes, all new wind turbines are normally equipped with lightning protection systems. The main functions of the lightning protection installed on a wind turbine are:

- Successful attachment/formation of the lightning strike to a preferred attachment point such as the air termination system on the blade.
- Facilitating the passage of the lightning current through the system into the earth without causing damage to systems including that damage that would result from high levels of electric and magnetic fields.
- Minimising levels of voltages and voltage gradients observed in and around the wind turbine.

These functions can be achieved by a number of different methods which have changed and improved with the development of new wind turbines. As they are particularly prone to lightning attachment owing to being the highest part of the turbine, the blades are usually considered an especially important part of the wind turbine in terms of lightning protection. The different types of lightning protection installed in wind turbines blades are:

- air termination systems on the blade surfaces
- high resistive tapes and diverters
- down conductors placed inside the blade
- conducting materials for the blade surface.

In all types of lightning protection systems (LPS), the metallic air terminations, strips and diverters and down conductors should be of sufficient cross-sections that they safely conduct the lightning current without any physical damage.

A system that is widely used is the internal lightning protection system consisting of an internal lightning down conductor capable of carrying the lightning current. Metal receptors, illustrated in Figure 2.22, which act as air terminations, penetrate the blade surface and are then connected to the down conductor. This system of external receptors connected to an internal down conductor is widely used for blades up to 60 m and the system is not likely to change for blades of larger lengths.

Wind turbine blades are subjected to high ambient electric fields during thunderstorms. In such scenarios areas of high electric field enhancement are prone to emit lightning leaders/answering leaders. In such cases it is important that these areas of high field enhancement are those which are connected to the earthing system and are capable of carrying the lightning current. The area of field enhancement depends on the geometrical shape and position of the structure. Most megawatt blades in service today have in-built lightning attachment systems. This report is mainly based on those with a receptor and a lightning down conductor. The lightning protection consists of exterior copper "receptor" air termination points, which are fastened to aluminium conductors running the length of the blade. This system is widely popular and is being used for wind turbines blades of

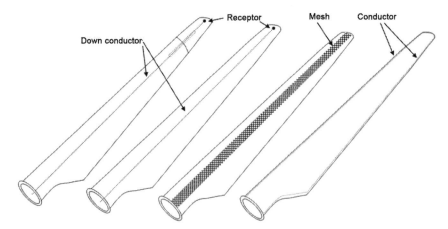

Figure 2.22 Lightning protection methods of wind turbine blades

up to 60 metres in length. The receptors are the points of high field enhancement and also the preferred location for lightning attachment. The number of receptors on the wind turbine blade depends on the manufacturer. The simulation shown in Figure 2.23 shows the behaviour of a wind turbine blade when subjected to an electric field, similar to that seen under a thunderstorm cloud.

Figure 2.23 shows a case where the receptors are the points of high electric field and thus theoretically the points for initial streamer growth, turning later into lightning leaders/answering leaders. The efficiency of the lightning protection system is dependent on the effectiveness of the air termination points. Once lightning successfully attaches to the wind turbine, it needs to be safely conducted away from the blade. This is done by the copper down conductor (in the case of the blade model in question) which runs through the entire length of the blade. It is important that the conductor itself is capable of carrying the rated lightning current without any damage. The size of the copper conductor is chosen with the help of the guidelines provided in IEC Standards (2010).

Even though extensive lightning protection measures are used there are still cases of lightning damage to wind turbines. Statistics of lightning damage to wind turbines has been covered in different parts of the literature. Some of the most widely referred to are given in Rademakers (2002). The IEC TR 61400-24 'Lightning Protection of Wind Turbines' is one of the most comprehensive studies regarding lightning protection of wind turbines. Initially developed as a technical report, it is now been reviewed as a full standard.

A summary of some of the main points of these surveys is given below (Rademakers 2002).

- 914 lightning damages have been reported over 11,364 operational years.
- 4 to 8 lightning faults have been experienced per 100 turbine years for northern European countries.
- 25 per cent of incidents are direct lightning strikes.
- 914 lightning damages represent only approximately 4 per cent of the total number of reported damages.

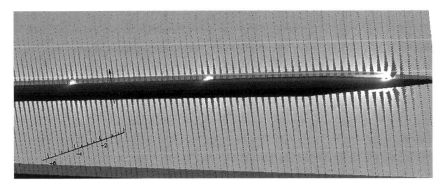

Figure 2.23 High field enhancement areas of a wind turbine blade when subjected to an electric field

As compared with the smaller wind turbines, the larger and newer wind turbines show fewer failures in the control system – a sign that the new wind turbines have a much more effective lightning protection system.

In a study performed by the National Renewable Energy Association in 2002, it is said that up to 8 out of 10 wind turbines could be expected to receive one direct lightning strike per year (Bruce 2007).

Another report from Germany stated that between 1992 and 1995, almost 124 direct lightning strikes were reported on wind turbines. Figure 2.24 shows that the control system is the part most affected by lightning strikes.

It must be noted that the control system is the most damaged part of the wind turbine. But while assessing the damage done to the wind turbine, the financial losses incurred while replacing a damaged part and also the related downtime must be taken into account. According to Biggs (2006), blade damage is the most expensive type of damage to repair. The rotor blades incur the highest losses due to downtime as well. A breakdown of the cost analysis in repairing individual components is given in Chafer's "The MOD wind energy safeguarding process" speech in 2007. This expense increases with the increase in the size of the wind turbine and also where it has been installed. Offshore repairs are obviously higher compared with those onshore.

RAM treatment and lightning protection

To reduce the wind turbine RCS different measures can be taken for different parts of the structure – for example careful shaping of the tower and nacelle and by applying RAM to the blade. The application of RAM on blades is considered a viable solution and prototypes have been fitted on commercial turbines. One of the challenges facing the treatment of the blades with RAM is the potential degradation of the lightning protection system as discussed earlier. The efficiency of the receptor-based lightning protection system is reduced if the blade is fully treated with RAM. Therefore a solution with partial RAM treatment is needed.

The RCS modelling presented here assumes no blade tilt or bending and is calculated by varying the yaw angle around a single blade at rotation angle of zero as

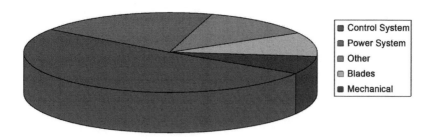

Figure 2.24 Lightning damage to different wind turbine components

shown in Figure 2.25. This orientation of the blade is chosen so that the maximum RCS level can be identified at approximately 180° and 360° yaw angles when the suction and pressure sides are illuminated respectively. In reality the RCS of a blade may be lower due to wind load bending, pitch and tilt angles. However, it is important to identify the peak RCS levels as these are the causes of "blade flashes" in the radar return, which can have a significant influence on the radar. Also, all RCS calculations are compared with a baseline which assumes that the untreated blade is made of a carbon fibre (CF) shell. Finally, since some RAM solutions are designed to absorb at narrow band of frequencies, the model was used to predict the RCS at 3 GHz. This approximates to the frequencies used by air traffic control radars.

This section shows the RCS reduction that can be achieved through RAM treatment of the blade surface while leaving a clearance area around the lightning receptor. Figure 2.25 shows a typical 40 m blade with three lightning receptors placed at 20 m (LR3), 30 m (LR2) and at the tip of the blade (LR1). It also shows the clearance radius, R_c, of the untreated area around each receptor. By varying R_c, the total treated area changes, which then affects the total RCS of the blade. Figure 2.26 shows RCS modelling results for an untreated blade, a fully treated blade and a blade with three receptors and R_c of 1 m.

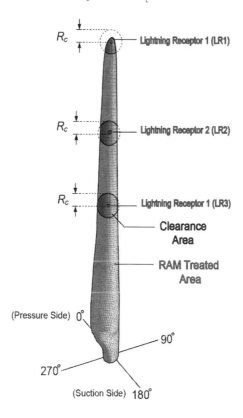

Figure 2.25 RAM treated blade with clearance area around the lightning receptors

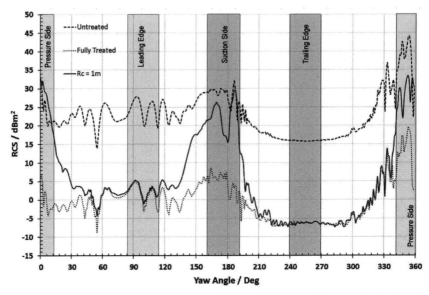

Figure 2.26 RCS comparison of an untreated blade, a fully treated blade and a blade with three lightning receptors and clearance radius $R_c = 1$ m

The results show that both fully treating the blade with RAM and leaving a clearance of 1 m around receptors give good reduction in RCS in the regions near the leading and trailing edges. When the leading and trailing edges of the blade are illuminated, the geometrical surface area of the first 35 per cent of the blade root make up more than 90 per cent of the total illuminated blade surface area. From the RCS view point, the blade root has large radii of curvature making it a very reflective portion of the blade. As the root portion of the blade is fully covered for all the clearance radii simulated, a significant reduction in the RCS is achieved when illuminating the leading and trailing edges.

Varying the clearance area around lightning receptors

When addressing the peak RCS regions, namely the suction and pressure sides, the RCS of the fully treated blade showed good reduction while introducing the 1 m clearance reduced the peak RCS by 12.2 dBm² and 2.6 dBm² in the pressure and suction sides respectively. The reduction in the total blade RCS is dependent on the size of the clearance area. Figure 2.27 gives a summary of RCS modelling showing the peak RCS level when illuminating the suction side, the trailing edge, the pressure side and the leading edge for different R_c values. The peak value for each orientation is taken as the maximum RCS value in the shaded region illustrated in Figure 2.26. The horizontal axis shows the variation of the clearance radius R_c from 0 to 3.5 m, where 0 represents a fully treated blade while the shaded region is the RCS of the untreated blade.

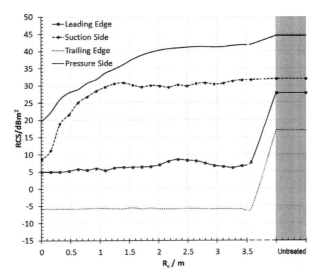

Figure 2.27 Peak RCS levels at the suction side, the trailing edge, the pressure side and the leading edge of RAM treated blade with three lightning receptors and varying R_c

The RCS of the trailing and leading edges remains relatively unchanged for all values of R_c due to the RAM treatment of the blade root section. On the other hand, the RCS of the suction and pressure sides rises significantly as R_c increases to 1.5 m. For R_c values larger than 1.5 m the peak RCS at the suction and pressure sides becomes close to that of the untreated blade. Lightning protection requirements may demand larger clearance areas around the receptors which will reduce the effectiveness of the RAM treatment. For three-bladed turbines within a large wind farm, this may still cause degradation in the performance of nearby radar systems. Hence, further reductions might be required. This may be achieved by using different R_c values for each receptor along the blade, or using fewer receptors. This is a question of detailed lightning protection design and may be subjected to lab testing using the IEC standards for lightning protection testing prior to implementation.

RAM vs blade tip RCS

ATC and other safety critical radar systems rely on the Doppler signatures of radar returns to identify targets of interest. The returns from fast-moving objects are processed while static or slow objects are suppressed, typically by the Moving Target Indicator (MTI) filter. Wind turbines with their large rotating blades are designed to track and face the flow of wind. Depending on the operation conditions and the orientation with respect to the radar, the blades' rotation causes the tips to move at high speeds causing a high Doppler shift in the radar returns. This may cause the returns from the blades to break through the Doppler and MTI filters, causing detection and initiation of false tracks or alteration of existing tracks.

The current generation of horizontal axis wind turbines has a distinctive blade profile. Although the exact geometry of the blade may vary depending on the manufacturer, a taper between the wide blade root and the very narrow tip is a norm. This means that the radii of curvature become increasingly small towards the tip of the blade. For RAM solutions with high angular sensitivity, the performance will gradually degrade as the curvature becomes steeper towards the tip. Hence, covering the very tip of the blade with RAM may have limited effect on the RCS of the leading and trailing edges near the blade tip. This can be seen when modelling only the last 6 metres of the blade and varying the clearance areas around the tip receptor as shown in Figure 2.28.

As seen earlier, the peak RCS of the tip can be reduced by applying RAM and the reduction is dependent on the size clearance patch. These peak values occur when the front and the back of the blade are illuminated. In such angles of illumination, the blade may have high RCS but produces a low Doppler shift. For the parts of the blade that may result in high Doppler, i.e., the leading and trailing edges, it can be noted that a fully treated blade tip reduced the RCS of the leading edge by only 16 dB rather than the specified 25 dB reduction of the RAM. Partial treatment of the tip by varying R_c shows that the leading edge RCS slowly rises to −5 dBm^2 and −2 dBm^2 at an R_c of 2.5 m and 3 m respectively. The RCS of the trailing edge remains relatively unchanged regardless of the size of the clearance patch. The RAM treatment of the sharp sections of the blade appears to have little effect on the RCS due to the RAM's angular sensitivity. However, this should not deter from considering RAM treatment as a potential solution. For this particular blade geometry the RCS of the tip's leading edge is 0.3 m^2 and 0.6 m^2 at R_c of 2 m and 3 m respectively. Depending on the radar system, application and subject to further studies, this might be considered to be small enough to be ignored at extended range.

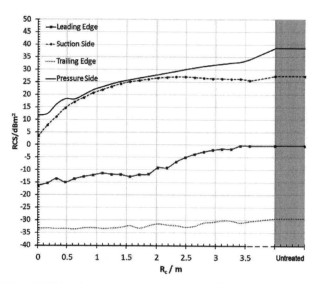

Figure 2.28 Peak RCS levels at the suction side, the trailing edge, the pressure side and the leading edge of RAM treated 6 m blade tip with varying R_c

References

Berger, K., 1967. Novel Observations on Lightning Discharges: Results of Research on Mount San Salvatore. *Journal of the Franklin Institute*, 283(6): pp. 478–525.

Berger, K., R. Anderson, and H. Kroninger, 1975. Parameters of lightning flashes. *Cigre' Electra*, 41: pp. 23–37.

Biggs, G., 2006, Mitigating the Adverse Effects of Wind Farms on an ATC Radar, in *BWEA28* 2006: Glasgow.

Bruce, G., 2007. *Effective Lightning Protection For Wind Turbine Generators*. Energy conversions, IEEE transactions on, 2007. 22(1): pp. 214–222.

Butler, M.M., and D.A. Johnson, 2003. Feasibility of Mitigating the Effects of Wind farms on Primary Radar, AMS Ltd, ETSU W/14/00623/REP.

Chafer, J., 2007. *The MOD wind energy safeguarding process*, All-Energy Conference Aberdeen, May 2007.

Cotton, I., 1998. *Lightning Protection of Wind Turbines*. PHD thesis: Solihull.

Howard, M. and C. Brown, 2004. Results of the electromagnetic investigations and assessments of marine radar, communications and position fixing systems undertaken at the North Hoyle wind farm by QinetiQ and the Maritime and Coastguard Agency. MCA Report MNA 53/10/366.

IEC Standards 61400–24, *"Wind turbine generator systems – Part 24: Lightning protection for wind turbines"* June 2010.

Marico, 2007. Investigation of Technical and Operational Effects on Marine Radar Close to Kentish Flats Offshore Wind Farm, BWEA Report.

Pinto, J. et al., 2006. Requirements Capture Summary Report. Stealth Technology for Wind Turbines: TP/2/RT/6/I/10117 APPS2B.

Poupart, G. J., 2003, Wind Farms Impact on Radar Aviation Interests, in DTI report number W/14/00614/00/REP, BWEA Radar Aviation Interests.

Rademakers, L., et al., 2002. *Lightning Damage of OWECS*. 2002(Part 1: "Parameters Relevant for Cost Modelling").

Rashid, L.S. and A.K. Brown, 2007. RCS and radar propagation near offshore wind farms. in Antennas and Propagation Society International Symposium, *2007 IEEE*.

3 Reliability and condition monitoring

Donatella Zappalá, Peter J. Tavner and Christopher J. Crabtree

Introduction

Condition monitoring (CM) allows early detection of any degeneration in system components, facilitating a proper asset management decision, improving availability, minimising downtime and maximising productivity. The adoption of cost-effective CM techniques plays a crucial role in minimising wind turbine (WT) operations and maintenance (O&M) costs for a competitive development of wind energy. As offshore WTs operate in remote locations and harsh environments, the need for high reliability and low O&M costs is higher than for onshore applications. For these reasons the development of reliable Condition Monitoring Systems (CMSs) for WTs is essential to avoid catastrophic failures and to minimise costly corrective maintenance.

This chapter outlines the current knowledge in the field of WT CM. Firstly, recent published WT reliability studies are summarised and the subassemblies that are of most concern for O&M are identified. This is followed by a description of the state of the art in CM of WTs, looking at both new emerging techniques currently being researched and industry developed tools, with a particular focus on the economic benefits of CMs. Conclusions are drawn about current systems challenges and limitations.

Wind turbine reliability

WT reliability is a critical factor in the economic success of a wind energy project. Poor reliability directly affects the project's revenue stream through both increased O&M costs and reduced availability to generate power due to turbine downtime.

The principal objective of reliability analysis is to gain feedback for improving design by identifying weaknesses in parts and subassemblies. Reliability studies also play a key role in optimising WT maintenance strategy. The main factor for optimal preventive maintenance, both from technical and economical points of view, is the time period selection for inspection. Optimum time period selection could not be achieved without a comprehensive reliability and availability analysis resulting in a ranking of critical subassemblies. In this way, WT reliability data can be used to benchmark WT performance for organising and planning future O&M, particularly offshore.

Understanding the WT failure rates and downtimes is difficult not only because of the considerable range of designs and sizes that are now in service worldwide but also since studies are conducted independently under various operating conditions in different countries (Pinar Pérez et al., 2013).

Whereas the standardisation of data collection practices is well established in the oil and gas industry (ISO 14224:2006), the wind industry has not yet standardised its methods for reliability data collection. However, an early wind industry reliability study, Wissenschaftlichen Mess und Evaluierungsprogramm database (WMEP) in Germany, developed a prototype data collection system described in Faulstich et al. (2008). Based on WMEP and other work, the EU FP7 ReliaWind Consortium (Wilkinson, 2011) proposed a standard approach to WT taxonomy and data collection, catering specifically for larger wind farms (WFs) and making use of both automatic but filtered Supervisory Control and Data Acquisition (SCADA) data and maintainers' logs. The achievement of a standardisation of WT taxonomy and data collection is of paramount importance to facilitate the exchange of information between parties and to ensure that data can be compared in a useful engineering and management way.

Although modern WTs currently have a shorter design life compared with traditional steam and gas turbine generator systems, i.e. 20+ years and 40 years respectively, their failure rates have been estimated at about three times those of conventional generation technologies (Spinato et al., 2009). Failure rates of 1–3 failure(s)/turbine/year for stoppages ≥ 24 hours are common onshore (Tavner et al., 2007). Despite substantial improvements in recent years, the current reliability of onshore WTs is still inadequate for the harsher offshore environment (Wilkinson et al., 2006) where maintenance attendance times are to be reduced and availability raised. For onshore WF sites, high failure rates can be managed by a maintenance regime that provides regular and frequent attendance to WTs. This will be costly or impossible to sustain in remote offshore sites. Failure rates of 0.5 failure(s)/turbine/ year would be desirable offshore, where planned maintenance visits need to be kept at or below 1 per year, but are nowhere near this level yet (Spinato et al., 2009).

Detailed measurements of failure rates from offshore WTs have not yet been accumulated in statistically significant numbers in the public domain, therefore the available literature on WT reliability focuses essentially on publicly available onshore data. WTs constitute a highly specialised technology and because of the commercial relevance of their failure data, due to the important capital investment, and therefore risk, of their installation, operators and manufacturers are reluctant to disclose data about reliability or failure patterns. As a result the sources of information are restricted to a few publicly available databases, although there is a strong argument for WT operators to end this restrictive practice in order to improve economic performance.

Figure 3.1 summarises the analysis of reliability data from three large surveys of European onshore WTs over 13 years. Failure and downtime data have been categorised by WT subassembly. The WMEP (Hahn et al., 2007),

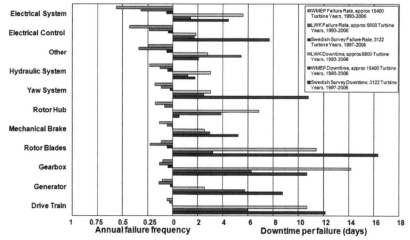

Figure 3.1 Failure rate and downtime results for onshore WTs from three large public domain reliability surveys (Ribrant and Bertling, 2007; Tavner et al., 2010)

the Landwirtschaftskammer (LWK) (Eggersglüß, 1995–2004) and the Swedish (Ribrant and Bertling, 2007) surveys provide large datasets of failure rate and downtime data which are remarkably similar and give valuable insights into the reliability of the various WT drive train subassemblies.

This data highlights that the highest failure rate subassemblies onshore do not necessarily cause the most downtime. Whilst electrical subassemblies appear to have higher failure rates and shorter downtimes, mechanical subassemblies, including blades, gearbox and generator components, tend to have relatively low failure rates but the longest downtimes. Similar results have been obtained by Pinar Pérez et al. (2013) who brought together and compared data from a selection of major reliability studies in the literature. From the failure rates the long downtime of the mechanical subassemblies is clearly not due to their intrinsic design weakness but rather the complex logistical and technical repair procedures in the field. It may result from the acquisition time for the spare part, i.e. supply chain, and for the required maintenance equipment. This will be aggravated particularly offshore where special lifting equipment and vessels are required and weather conditions have to be considered. In particular, the gearbox exhibits the highest downtime per failure among all the onshore WT subassemblies and it is the most critical for WT availability. Also, it has been shown that the replacement of WT major components, such as the gearbox, is responsible for 80 per cent of the cost of corrective maintenance (Besnard, 2013). This suggests that drive train subassemblies such as the generator and gearbox warrant the most attention.

The datasets shown in Figure 3.1 have some important limitations. Data are taken from mixed and changing WT populations. Many of the turbines are old and some of them may be in the wear out phase of the bath tub curve. Also, the majority of turbines are much lower power than modern turbines and use dated technologies.

The more recent ReliaWind study (Wilkinson et al., 2010) attempted to address these limitations by considering only turbines that met the following requirements: rating greater than or equal to 850 kW, variable speed, pitch regulated and operating for a minimum of two years; WFs with at least 15 turbines. Figure 3.2 and Figure 3.3 show the results of the ReliaWind study where the turbine was broken into a detailed taxonomy of subsystems and subassemblies to identify critical areas of interest from > 4000 onshore WT-years. Unlike the data shown in Figure 3.1, for reasons of confidentiality the published ReliaWind results do not show the actual failure rate and downtime, only the percentage distribution. In spite of the diverse technologies and power ratings, the ReliaWind findings are broadly comparable with the WMEP, LWK and Swedish surveys and the same failure rate trend emerges. However, the downtime trend shows much greater emphasis on the rotor and power modules because it is believed that these newer variable speed WTs have not yet experienced major gearbox, generator or blade failures to date in service (Tavner, 2012).

A recent study (Faulstich et al., 2011) has shown that onshore 75 per cent of WT failures are responsible for only 5 per cent of the downtime, whereas only 25 per cent of failures cause 95 per cent of downtime. Downtime onshore is dominated by a few large faults, many associated with gearboxes, generators and blades, requiring complex and costly repair procedures. The 75 per cent of faults causing 5 per cent of the downtime are mostly associated with electrical faults, often caused by the system tripping, which, in the majority of cases, are relatively easy to fix via remote or local resets in the onshore environment. However, as WTs go offshore, limited accessibility, longer waiting, travel and work times will amplify the influence of the 75 per cent short duration failures on offshore WT availability. Local resets will carry high costs and difficult access conditions that are likely to significantly increase the downtime contribution of these subassemblies (Tavner et al., 2010).

As mentioned, very little field data is publicly available for offshore WFs, although there are a number of reports published from early publicly funded projects in Europe (Feng et al., 2010). Crabtree (2012) carried out a reliability analysis of 3 years of available data from Egmond aan Zee offshore WF in the Netherlands (NoordzeeWind, 2007–2009). The WF consists of 36 Vestas V90–3 MW WTs situated 10–18 km offshore and in 17–23 m water depth in the North Sea. Operational reports, available for 2007 to 2009, gave the number of stops resulting from 13 subassemblies or tasks, representing 108 WT-years of data. The results are shown in Figure 3.4 in the same format as Figure 3.1. It should be noted that, in this case, stop and not failure frequency was recorded. Direct comparison cannot be made between the onshore and offshore datasets, Figure 3.1 and Figure 3.4 respectively, as stops and failure frequencies are different concepts. However, the overall distribution is largely similar with subassemblies with high stop and failure rates not always being the worst causes of downtime. The control system dominates the number of stops, 36 per cent, but caused only 9.5 per cent of the total downtime. Conversely, the gearbox and generator respectively contributed only 6.7 per cent and 2.8 per cent of total stops but 55 per cent and 15 per cent respectively of the downtime. The average energy lost per turbine per year demonstrates similar trends to the downtime (Crabtree, 2012).

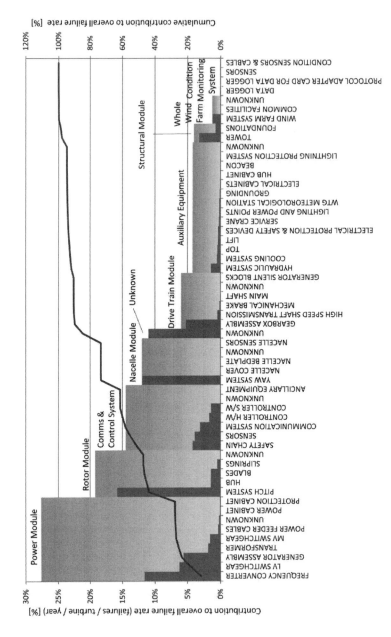

Figure 3.2 Distribution of normalised failure rate by sub-system and subassembly for WTs of multiple manufacturers from the ReliaWind survey (Wilkinson et al., 2010)

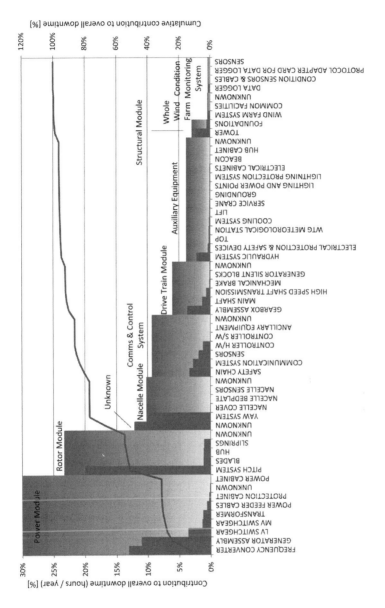

Figure 3.3 Distribution of normalised downtime by sub-system and subassembly for WTs of multiple manufacturers from the ReliaWind survey (Wilkinson et al., 2010)

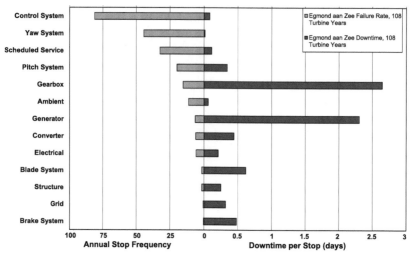

Figure 3.4 Stop rate and downtime data from Egmond aan Zee WF over 3 years (Crabtree, 2012)

For large, remote WFs to become cost effective, improving WT availability is essential and requires the adoption of a number of design and operational measures, for example the choice of the most effective turbine architecture, the installation of effective CM and the application of appropriate O&M programmes. Emphasis should be placed on avoiding large maintenance events that require deploying expensive and specialised equipment. However, the most important measure will be to improve the intrinsic reliability of the turbines used, achievable only through close collaboration between manufacturers, operators and research institutes, and the development of a standardised methodology for reliability data collection and analysis (Feng et al., 2010).

Wind turbine monitoring systems

The need for successfully detecting incipient faults before they develop into serious failures, to increase availability and lower cost of energy, has led to the development of a large number of WT monitoring systems. As the wind industry develops these monitoring systems are slowly being integrated together.

As described in Tavner (2012), modern WTs are equipped with monitoring systems which allow the active remote control of their functions. The WT monitoring may include a variety of systems as follows:

- Supervisory Control & Data Acquisition (SCADA) system
- Condition Monitoring System (CMS)
- Structural Health Monitoring (SHM).

The general layout and interaction of the various WT monitoring systems is shown in Figure 3.5.

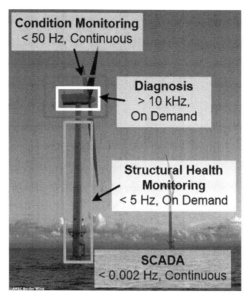

Figure 3.5 Structural health and condition monitoring of a wind turbine (Crabtree et al., 2014)

Supervisory Control & Data Acquisition (SCADA) system

The SCADA system is a standard installation on large WTs, their data being collected from individual WT controllers. The SCADA system provides low-resolution monitoring to supervise the WT operation and provide a channel for data and alarms from the WT. The status of the WT and its subassemblies is assessed using sensors fitted to the WT which measure meteorological and turbine operating information at a low data sample rate, usually at 10-minute intervals, in the WT controller. SCADA is a valuable low-cost monitoring system, integrating cheap, high-volume measurement, information and communication technology. Data is used to control the WT and transmitted to a central database for large WT original equipment manufacturers (OEM), however WF operators rarely use SCADA to monitor WT and WF performance. A recent trend in wind project management is to tie the SCADA data into centralised monitoring centres, operated by either the turbine manufacturer or a service provider. This arrangement can provide 24/7 coverage for multiple projects distributed over several time zones. The volume of data from many installations in varied conditions provides better visibility of recurrent problem causes and a synergistic sharing of information.

Commercially available SCADA systems

A survey of the commercial SCADA systems available to the wind industry, conducted by Durham University and the UK SUPERGEN Wind Energy Technologies Consortium, is given in Chen et al. (2014) and is provided in Appendix A. This survey provides an up-to-date insight, at the time of writing, into the current state of the art of the commercial SCADA Data Analysis Tools and shows the range of systems currently available to WT manufacturers and operators. The document contains information gathered over several years through interaction with SCADA monitoring system and turbine manufacturers and includes information obtained from various product brochures, technical documents and personal interaction with sales and technical personnel at the European Wind Energy Association (EWEA) Conferences from 2011 to 2014. The detailed table from the SUPERGEN Wind survey, listing the up-to-date commercially available SCADA systems for WTs, from which this summary is derived, is provided in Appendix A. The systems are grouped by monitoring technology and then alphabetically by product name.

The survey shows that most of the commercially available SCADA systems are able to analyse real-time data. The WT performance analysis techniques used vary from tailored statistical methods to the use of artificial intelligence. A number of commercial SCADA systems, such as Enercon SCADA System and Gamesa WindNet, are customisable Wind Farm Cluster Management Systems, providing, for both individual WTs and WFs, a framework for data acquisition, remote monitoring, open/closed loop control, alarm management, reporting and analysis, production forecasting and meteorological updates. Five out of the 26 products surveyed, GH SCADA, OneView SCADA system, SgurrTREND, WindHelm Portfolio Manager and Wind Turbine In-Service, were developed by renewable energy consultancies in collaboration with WT manufacturers, WF operators, developers and financiers to meet the needs of all those involved in WF operation, analysis and reporting.

Some of the SCADA systems surveyed, such as Alstom WindAccess, claim to feature built-in diagnostics techniques for a timely diagnosis of WT component failures. In these systems remotely collected data is used to establish benchmarks and identify irregularities, allowing timely intervention to avoid unplanned outages or secondary damage.

Recently proposed SCADA systems, such as the Wind Turbine Prognostics and Health Management platform developed by the American Center for Intelligent Maintenance Systems (IMS), feature WT modelling for predictive maintenance and a multi-regime diagnostic and prognostic approach to handle the WTs under various highly dynamic operating conditions.

Some recent SCADA solutions, such as the Mita-Teknik Gateway System and ABS Wind Turbine In-Service system, can be adapted and fully integrated with commercially available conventional vibration-based CMSs using standard protocols. These PC-based software packages are designed to collect, handle, analyse and illustrate both operational parameter data from the WT controller and CMS vibration signals/spectra with simple graphics and text. This unified plant operations' view allows a broad and complete analysis of the turbine's conditions by considering signals of the controller network as well as condition monitoring signals.

In some cases the SCADA product developer also offers service contracts for SCADA systems beyond the manufacturer service and warranty. They usually include hardware audit, a system-specific maintenance plan, monthly checks of the SCADA system and 24/7 online support. Examples are ABS Wind Turbine In-Service and SCADA International OneView SCADA system.

Examples of wind turbine monitoring through SCADA systems

A great benefit of SCADA is that it gives an overview of the whole WT by providing comprehensive signal information, historical alarms and detailed fault logs, as well as environmental and operational conditions. The main weaknesses of SCADA are that the large volume of data generated requires considerable analysis for on-line interpretation and that the low data rate does not allow the depth of analysis usually associated with accurate diagnosis. Potentially, SCADA alarms can help a turbine operator to understand the WT and key components status, but in a large WF, these alarms are currently too frequent for rational analysis. Their added values could be explored in more detail.

Recent researches, including Zaher et al. (2009), Gray and Watson (2010) and Chen et al. (2015), have shown how rigorous analysis of the information collected by SCADA systems can provide long-term fault detection, diagnosis and, in some cases, prognosis for the main WT drive train subassemblies, such as the gearbox, the converter and the pitch system.

Zaher et al. (2009) propose an automated analysis system for early fault identification of the main WT components based on Artificial Neural Networks (ANN). The results from 52 WT-years of data show that the proposed techniques can automatically interpret the large volumes of SCADA data presented to an operator and highlight only the important aspects of interest to them. This allows the use of routinely logged SCADA information for the identification of turbine faults by detection of anomalous performance measurements, dramatically reducing the information presented to the operators. The proposed multi-agent platform allows the techniques to be brought together to corroborate their output for more robust fault detection and also allows the development of a system that can be used to apply the techniques across a complete WF.

Gray and Watson (2010) present a methodology for continuous, online calculation of damage accumulation, and hence risk of failure, in a typical 3-stage WT gearbox design using standard turbine performance parameters and physics of failure approach. The proposed approach allows real-time accurate estimates of the probability of failure for specific failure modes and components and has been illustrated for a specific failure mode using a case study of 400 WT-years SCADA data of a large WF where a significant number of gearbox failures occurred within a short space of time. However, the main weakness of this methodology is that it requires an in-depth understanding of the dynamics of the gearbox under all kinds of conditions and loads.

Chen et al. (2015) propose a fault prognosis procedure using the a-priori knowledge-based Adaptive Neuro-Fuzzy Inference System (APK-ANFIS) to analyse WT SCADA data, concentrating particularly on WT electric pitch system faults. In order to construct the proposed system, the data of the six known pitch faults from two Spanish WTs were used to train the system with a-priori knowledge incorporated. The effectiveness of the approach was demonstrated by testing the trained system against a Spanish WF containing 26 WTs to demonstrate the prognosis of pitch faults. The method was then extended to 153

WTs of different design at Brazos in the USA using a different pitch technology. Results were then compared with an Alarm approach to demonstrate the advantage of the prognostic horizon. The prognosis results are presented in Figure 3.6. They show that the ANFIS approach gives significant warning of pitch faults with prognostic horizons up to 21 days. This is much better than the common SCADA alarm approach, which only gives prognostic horizons of a few days. The result of this research has demonstrated that the proposed APK-ANFIS approach has strong potential for online WT pitch fault prognosis.

Simple signal algorithms to analyse SCADA alarms and to detect WT component faults have been developed by Qiu et al. (2012) and Feng et al. (2013), respectively, although the techniques still need further verification.

Qiu et al. (2012) propose a time-sequence and Bayesian probability-based analysis to rationalise and reduce SCADA alarms considering data from two large populations of modern onshore WTs over a period of 1–2 years. The two methods of SCADA alarm analysis are then applied to the data from one of the WT populations to detect pitch and converter system known malfunctions, identified from fault logs. Figure 3.7 shows the alarm time-sequence results for the converter, with the timing linkages between the 12 alarms observed during the failure. The strength of the time-sequence method is its simplicity and clarity, in which it can be applied online, and its diagrams can be generated rapidly to show the linkages between alarms and errors, which is valuable for fault detection. However, this approach does not show alarm root cause. Figure 3.8 shows the results from the converter fault probability-based analysis. The probability of each SCADA alarm occurrence is approximately represented by the circle size, while the conditional probability of two alarms is approximately represented by the size of their corresponding overlapping areas. Figure 3.8 shows

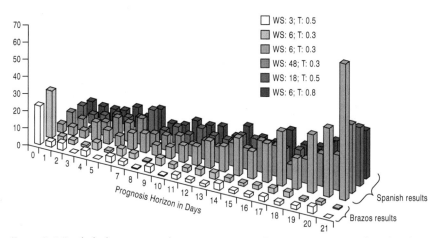

Figure 3.6 Pitch fault prognostic horizon comparison between two types of wind turbine in Spain (4 WTs) and Brazos USA (2 WTs) WS = Analysis Window Size, T = Analysis Threshold. Reproduced by permission of the Institution of Engineering & Technology (Chen et al., 2015)

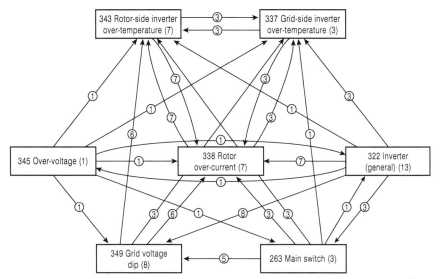

Figure 3.7 Converter Alarm Transition Time-sequence Analysis (Qiu et al., 2012)

Boxes show alarm codes and descriptions with number of actuations in brackets. Transition lines show number of transitions from one alarm to another.

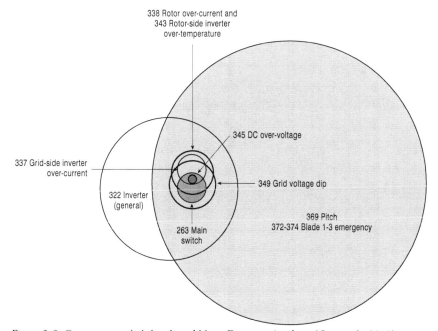

Figure 3.8 Converter probability-based Venn Diagram Analysis (Qiu et al., 2012)

Venn circles show alarm codes & descriptions. Grey highlighted Venn circle indicates root cause.

a process where grid voltage dips cause rotor-side inverter over-currents, DC link over-voltages, which consequently cause grid-side inverter over-currents and grid-side inverter over-temperatures. The DC link over-voltage is a key trigger for subsequent alarms. The probability-based approach shows clearly the alarm root cause, however it requires a significant time period analysis in order to provide good quality results. Both methods have the potential to systematically optimise WT alarm analysis, providing valuable fault detection, diagnosis and prognosis from the conditions under which the alarms are generated, with the ultimate aim of improve WT reliability.

Feng et al. (2013) apply basic thermodynamics laws to the gearbox to derive robust relationships between temperature, efficiency, rotational speed and power output. The retrospective application of these relationships to a case study has shown the successful detection and prediction of a gearbox planetary stage failure in a working 2 MW class variable-speed WT by monitoring the machine power output and rotational speed, and relating them to the SCADA gearbox oil temperature rise. The results presented in figures 3.9 and 3.10 clearly show a rising gearbox oil temperature in the 9 months before failure, with a worsening trend presented 3 months before failure, in response to a turbine efficiency reduction due to the development of the failure. These results clearly demonstrate how slow speed SCADA data is perfectly suitable for the long-term detection of gearbox problems.

The implementation of the proposed techniques in the field would result in more intelligent approaches for SCADA data analysis for automatic WT fault diagnosis and prognosis, giving the operators sufficient time to make more informed decisions regarding the asset maintenance.

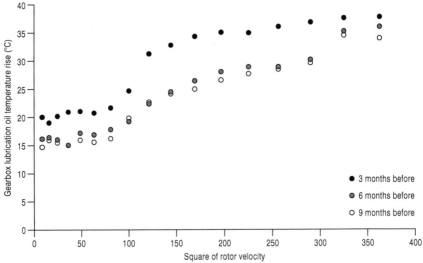

Figure 3.9 WT Gearbox oil temperature rise trend versus square of rotor velocity different number of months before gearbox failure (Feng et al., 2013)

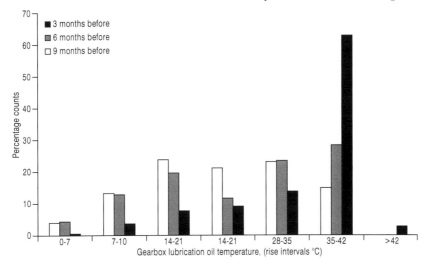

Figure 3.10 WT gearbox oil temperature rise histogram of occurrences, demonstrating clear deterioration in the number of months before gearbox failure (Feng et al., 2013)

Structural Health Monitoring (SHM)

SHM provides low-resolution signals for the monitoring of key items of the WT structure, tower and foundations. These are particularly important offshore where the structures are subjected to strong effects from the sea, wind and seabed. Structural faults are slow to develop and do not need continuous monitoring. SHM systems are frequently installed below the nacelle on large WTs, i.e. > 2 MW. Low-frequency sampling signals, below 5 Hz, are recorded from accelerometers or similar low-frequency transducers to determine the structural integrity of the WT tower and foundation for faults driven by blade-passing frequencies, wind gusts and wave slam. A detailed review of new emerging techniques currently being researched in the SHM field is provided in Ciang et al. (2008).

Condition Monitoring System (CMS)

CMSs provide high-resolution monitoring of WT high-risk subassemblies for the diagnosis and prognosis of faults. Included in this area are Blade Monitoring Systems (BMSs), aimed at the early detection of blade defects.

CMS *state of the art*

Condition monitoring focuses on remotely measuring critical indicators of WT component health and performance, with the objective of identifying incipient failures before catastrophic damage occurs. Failures of the major components of the WT drive train are expensive due to the high costs for the spare parts, the logistic and maintenance equipment, and energy production losses. A comprehensive on-line monitoring programme provides diagnostic information on the health of the turbine subsystems and alerts the maintenance staff to trends that may be developing into failures or critical malfunctions. This information can be used to schedule maintenance tasks or repairs before the problem escalates and results in a major failure or consequential damage with resultant downtime and lost revenue. Thus necessary actions can be planned in time and need not be taken immediately; this factor is of special importance for offshore plants where bad weather conditions can prevent any repair actions for several weeks. Many faults can be detected while the defective component is still operational. In some cases, remedial action can be planned to mitigate the problem. In other cases, measures can be implemented to track the problem's progression. In the worst case of an impending major failure, CM can assist maintenance staff in logistics planning to optimise manpower and equipment usage and to minimise the cost of a repair or replacement. In this condition, both the cost of the maintenance activity itself and the costs for production losses may be reduced.

CM has become increasingly important in larger turbines owing to the greater cost of the components and greater concern about their reliability. Furthermore, it represents an essential technology for offshore turbines, due to their projected size, limited accessibility and consequent need for greater reliability. The main advantages and benefits arising from applying CM are summarised in Table 3.1 (Hameed et al., 2009).

The processes necessary for a successful CM approach are:

- Detection: the essential knowledge that a fault condition exists in a machinery component and ideally its location. Without this, no preventive action can be taken to avoid possible system failure.
- Diagnosis: the determination of the nature of the fault, including its more precise location. This knowledge can be used to decide the severity of the fault and what preventive or remedial action needs to be taken (if any).
- Prognosis: the forecast or prediction of the remaining life or time before failure. Based on this, the most efficient and effective action to remove the

Table 3.1 Characteristics of CMSs, adapted from (Hameed et al., 2009)

Characteristics	Advantages	Benefits
Early warning	Avoid breakdowns Better planning of maintenance	Avoid repair costs Minimise downtime
Identification of problem	Right service at the right time Minimising unnecessary replacements Problems resolved before the time of guarantee expires	Prolonged lifetime Lowered maintenance costs Quality-controlled operations during time of guarantee
Continuous monitoring	Constant information that the wind power system is working	Security, less stress

fault can be planned. Prognosis has the largest potential payoff of all CM technologies, especially for offshore turbines.
* Maintenance action: repair or replacement actions to remove the cause of the fault.

The level of detail required for failure prevention depends very much on the type of component, its perceived value and the consequences of failure. For the main and most expensive WT subassemblies, such as the gearbox, the generator and the blades, simple fault detection is not sufficient, as their cost is usually too high to justify total replacement, and some form of diagnosis is required. Diagnosis also allows scope for prognosis, either to predict the possibility of progression from a non-critical fault to a critical fault, indicating the need to closely monitor the fault progression, or to predict the time to failure, allowing scheduling of repairs.

Review of CMS techniques

On-line monitoring and fault detection are relatively new concepts in the wind industry and are flourishing on a rapid scale. Today it is the state of the art for modern onshore and offshore WTs to be equipped with some form of CMS. Measurements are recorded from sensors and different methodologies and algorithms have been developed to analyse the data with the aim of monitoring the WT performance and identifying characteristic fault indicators.

Poor early reliabilities for gearbox and drive train components have led to an emphasis on WT CM of drive train components. CM is believed to have entered the WT market about twenty years ago as a result of a series of catastrophic gearbox failures in onshore turbines, which led to insurers demanding that WT manufacturers take remedial action by utilising CM technology, already applied to other rotating machinery. However, the more recent information on WT reliability and downtime, shown in Figure 3.2 and Figure 3.3, suggests that the target for CM should be widened from the drive train towards WT electrical and control systems and blades.

In recent years, efforts have been made to develop efficient and cost-effective CM techniques and signal processing methods for WTs. There have been several reviews on WT CM in literature, including Amirat et al. (2007), Garcia Marquez et al. (2012), Hameed et al. (2009), Hyers et al. (2006), Lu et al. (2009), Wiggelinkhuizen et al. (2008), Wilkinson et al. (2007) and Yang et al. (2014), discussing the main CM techniques, the signal processing methods proposed for fault detection and diagnosis and their applications to wind power.

For advanced CM approaches, signal processing techniques are used to extract features of interest. The selection of appropriate signal processing and data analysis techniques is crucial to WT CM success. If the fault related characteristics can be correctly extracted using these techniques, fault growth can be assessed by observing characteristic variations. These characteristics are also important clues for fault diagnosis.

The most recent reviews by Yang et al. (2014) and Garcia Marquez et al. (2012) give a particularly insightful and detailed summary of the state of the art in WT CMSs while also providing a comprehensive explanation of the new emerging techniques currently being researched.

The following monitoring techniques, available from different applications and possibly applicable for WTs, have been identified:

1   Vibration analysis
2   Oil analysis
3   Strain measurements
4   Thermography
5   Acoustic emissions
6   Electrical signals.

Among these different techniques, vibration analysis and oil monitoring are the most predominantly used for WT applications due to their established successes in other industries.

Vibration analysis

Vibration analysis is a low-cost and a well-proven monitoring technology typically used to monitor the condition of WT rotating components, i.e. the drive train. Vibration techniques were the first to be used in WT CMS. The principle is based on two basic facts (Randall, 2011):

1   Each component of the drive train has a natural vibration frequency and its amplitude will remain constant under normal conditions, although varying with drive train speed.
2   The vibration signature will change if a component is deteriorating and the changes will depend on the failure mode.

The type of sensors used essentially depends on the frequency range of interest and the signal level involved. A variety of techniques have been used including low-frequency accelerometers for the main bearings and higher-frequency accelerometers for the gearbox and generator bearings and in some cases proximeters. By far the most common transducer in use today is the piezoelectric accelerometer, which is applicable to a broad range of frequencies, is inexpensive, robust, and available in a wide range of sizes and configurations.

Principles for vibration analysis are presented in detail in McGowin et al. (2006). Signal analysis usually requires specialised knowledge. Almost all of the commonly used algorithms can be classified into two categories: time and frequency domains. Time domain analysis focuses principally on statistical characteristics of vibration signal such as peak level, standard deviation, skewness, kurtosis, and crest factor. Frequency domain analysis uses Fourier methods, usually in the form of a fast Fourier transform (FFT) algorithm, to transform the time domain signal to the frequency domain. This is the most popular processing technique for vibration analysis. Further analysis is usually carried out conventionally using vibration amplitude and power spectra. FFT analysers provide constant bandwidth on a linear frequency scale, and, by means of zoom or extended lines of resolution, they also provide very high resolution in any frequency range of interest. This permits early recognition and separation of harmonic patterns or sideband patterns and separation of closely spaced individual components. The advantage of frequency domain analysis over time domain analysis is its ability to easily identify and isolate certain frequency components of interest.

Applying vibration based CM to WTs presents a few unique challenges. WTs are variable load and speed systems operating under highly dynamic conditions, usually remote from technical support. This results in CM signals that are dependent not only on component integrity but also on the operating conditions. One limitation of the conventional FFT analysis is its inability to handle non-stationary waveform signals that may not yield accurate and clear component features. To overcome the problems of the conventional FFT-based techniques and find improved solutions for WT CM, a number of advanced signal processing methods, including wavelet transforms, that enable to detect defects whose vibration signature is not cyclic, time–frequency analysis, bi- or tri-spectrum and artificial intelligence techniques have been researched recently (Amirat et al., 2009, Garcia Marquez et al., 2012, Hameed et al., 2009, Yang et al., 2009b, 2010, 2014, Zaher et al., 2009). However, the interpretation of the results is more complicated for direct analysis than FFT techniques. Moreover, most new techniques are unsuitable for on-line CM use because they are computing intensive and have not been demonstrated yet in the operating WTs.

Oil analysis

Oil analysis is used to determine the chemical properties and content of oil coolant or lubricant with two purposes: safeguarding the oil quality and safeguarding the components involved. Although still expensive, on-line debris detection in lubricant oil is one of the most promising techniques for use in WT CM. Oil analysis focuses on one of the most critical WT components, the gearbox. Gear wheels and bearings deterioration depends mainly on the lubricant quality, i.e. particle contamination and properties of the oil and additives used to improve the performance of the oil. Oil monitoring can help detect lubricant, gear and bearing failures and is an important factor in achieving maximum service life for WT gearboxes. Oil analysis is discussed in detail in McGowin et al. (2006). On-line oil analysis is gradually becoming more important with several on-going pilot projects. Crucial to the value of oil debris detection is the length of the warning that it can give of impending failure, giving time to arrange for inspection and maintenance (Tavner, 2012). Little or no vibration may be evident while faults are developing, but analysis of the oil can provide early warnings. However, oil debris detection cannot locate a fault, except by distinguishing between the types of debris produced. The combined use of vibration and oil analysis, to cover a broader range of potential failures and to increase the credibility of the CM results, could be a key to WT drive train monitoring.

Strain measurements

Strain measurements by fibre-optic sensors are proving a valuable technique for measuring blade-root bending moments as an input to advanced pitch controllers and can be used to monitor WT blades. They have been demonstrated in operation, however they are still too expensive and improvements in costs and reliability are expected. The use of mechanical strain gauges can be very useful for lifetime forecasting and protecting against high stress levels, especially in the blades. However, mechanical strain gauges are not robust on a long term, as they are prone to failure under impact and fatigue load. Developments are still necessary for the improvement of instrumentation and sensor robustness as well as for reliable fault detection algorithms.

Thermography

Thermography is often used for monitoring electronic and electric components and identifying failure. Infrared thermography is a technique used to capture thermal images of components. Every object emits infrared radiation according to its temperature and its emissivity. The radiation is captured by a thermographic camera. The technique can be applied to equipment from time to time but should be done when the equipment is fully loaded, and often involves visual interpretation of hot spots that arise due to bad contact or a system fault. At present the technique is not particularly well established for on-line CM, but cameras and diagnostic software that are suitable for on-line process monitoring are starting to become available.

Acoustic emissions

Acoustic emissions could be helpful for detecting drive train, blade or tower defects. Rapid release of strain energy takes place and elastic waves are generated when the structure of a metal is altered, and this can be analysed by acoustic emissions. Acoustic monitoring has some similarities with vibration monitoring but also a principal difference. Whereas vibration sensors are mounted on the component involved, so as to detect movement, acoustic sensors are attached to the component by flexible glue with low attenuation. For vibration analysis the frequencies related to the rotational speeds of the components are of interest. For acoustic emission a wider bandwidth of higher frequencies is considered which can give an indication of starting defects. These sensors have lately gained much more attention in order to detect early faults and are successfully applied for monitoring bearings and gearboxes. Acoustic emission is considered more robust for low-speed operation of WT compared with the classic vibration based methods. However, this technique is still too expensive due to the data acquisition costs.

Electrical signals

Electrical signals have been widely used for CM of rotating electric machines and their coupled drive trains (Tavner et al., 2008) but have not been used in WT because of lack of industry experience. Voltage, current and power measurements, used to control the generator speed and excitation, represent the newest potential source of CM information for WTs. The difficulty with these electrical signals is that they are very rich in harmonic electrical information, which must be accurately understood if diagnosis is to be performed with confidence (Yang et al., 2009b). These techniques are at the moment confined to research-related activities but there is significant potential for applying them successfully in the field (Djurović et al., 2012). Work by Watson et al. (2010) showed the potentiality of the wavelet transform in detecting WT mechanical and electrical faults by electrical power analysis but this technique still has strong practical limitations because of its intensive calculation and consequent inefficiency in dealing with lengthy WT signals.

Some of the more recent emerging CM techniques described in the research literature (Yang et al., 2014, Garcia Marquez et al., 2012) include ultrasonic testing, potentially effective for detecting early WT blade or tower defects, shaft torque and torsional vibration measurements, for main shaft and gearbox monitoring, and the shock pulse method, an on-line approach to detecting WT bearing faults.

To date, little work has been done in the area of prognosis models. Much of the research in this area is generic and being conducted by the aerospace community for civil and military aviation. Some specific research will be required to apply the principles of prognosis to the wind power industry (Hyers et al., 2006). A model for estimating the residual lifetime of generator bearing failure based on CMS data has been recently proposed in Fischer (2012).

Commercially available CMSs

The WT CMS application was requested by insurance companies in Europe, such as Germanischer Lloyd (2007), in the late 1990s following a large number of claims triggered by catastrophic gearbox failures (Tavner, 2012). As the drive train is not only one of the most valuable WT subsystems but also one of the most trouble prone, German insurers introduced this clause as a cost deterrent to encourage an improvement in its operating life.

A typical CMS should feature:

- Physical measurement: sensors measure the machine signals, analogue or pulse signals from sensors are then filtered and converted to digital information;
- Data Acquisition System (DAS): data is transmitted from sensors to processing unit;
- Feature extraction: characteristic information is extracted from raw sensor data;
- Pattern classification: defect type and severity level are diagnosed;
- Life prediction: remaining service life of the monitored component is prognosticated.

Today, a number of commercial certified WT CMSs are available to the wind industry and they are largely based upon the successful experience of monitoring conventional rotating machines which is beginning to adapt to the WT environment. A survey of the commercially available WT CMSs, conducted by Durham University and the UK SUPERGEN Wind Energy Technologies Consortium, is given in Crabtree et al. (2014) and in Appendix B. This survey provides an up-to-date insight, at the time of writing, into the current state of the art of CM and shows the range of systems currently available to WT manufacturers and operators. The document contains information gathered over several years through interaction with monitoring system and turbine manufacturers and includes information obtained from various product brochures, technical documents and personal interaction with sales and technical personnel at the European Wind Energy Association Conferences from 2008 to 2014. The detailed table from the SUPERGEN Wind survey, listing the up-to-date commercially available CMSs for WTs, from which this summary is derived, is provided in Appendix B. The systems are grouped by monitoring technology and then alphabetically by product name.

The survey shows that the large majority of CMSs currently in use on operational WTs are based on vibration monitoring of the drive train, at a typical sampling rate up to 20 kHz, with special focus on main bearing, gear teeth and bearings. Figure 3.11 provides an example setup for a vibration-based WT CMS (Doner, 2009). The system consists of several sensors and a DAS, referred to as the condition diagnostics system (CDS) enclosure, which are located in the turbine nacelle, and a data server, CDS server, located at the wind park or a remote monitoring centre. Typically, the DAS has a channel for the shaft rotational speed signal, which is either measured by a tachometer dedicated to the CM system or supplied by the turbine controller. The communication between the DAS and the data server located at the wind park can be through Ethernet or fibre optic cables. If no data server is set up at the local wind park, the DAS normally can be configured to wirelessly transmit the test data to a server located in the remote monitoring centre, which could be anywhere around the globe. The data server normally hosts the CM software package, which is a platform for reviewing and analysing the data, presenting the CM results, and streamlining both raw and processed data into a CM database. One wind park, typically consisting of hundreds of turbines, can be monitored by one CM software package.

Among vibration-based CM systems, the main differences are the number of sensors, measurement locations, and analysis algorithms, since almost all systems use standard accelerometers as the main physical measurement device. The typical configuration of the sensors along the WT drive train is shown in Figure 3.12, which refers to the Gram & Juhl Turbine Condition Monitoring (TCM) system architecture (Gram & Juhl A/S, 2010). Sensors are mounted on the bearing housing or gear case to detect characteristic vibration signatures for each component. The signature for each gear mesh or bearing is unique and depends on the geometry, load, and speed of the components.

According to the survey, the most popular signal processing technique for the vibration monitoring of the WT drive train is the traditional FFT analysis of the high frequency data to detect the fault-specific frequencies. Frequently, time domain parameters are used to monitor the trend of overall vibration level over time. To minimise data transmission, CMSs analyse data and transmit trends to the system microprocessor continuously, whereas spectral analysis occurs only

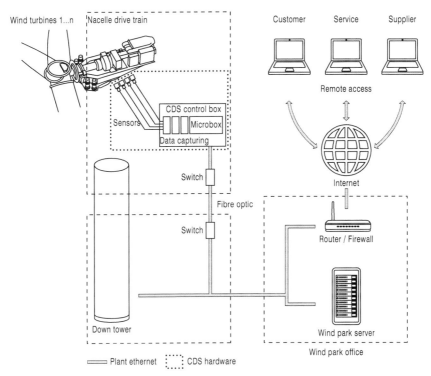

Figure 3.11 Example setup of a WT CMS, adapted from the Winergy CDS (Doner, 2009)

when settings detect an unusual condition. One triggering mechanism, such as time interval-based or vibration level-based, can be set up in the time domain parameter overall trending process. Whenever it triggers, a discrete frequency analysis snapshot can be taken. Based on these snapshots, detailed examinations of the component health can be conducted. The amplitude of characteristic frequencies for gears, e.g., meshing frequency, and bearings, e.g., ball passing frequency, can also be trended over time for detecting potential failures. Such a strategy mitigates the burden of data transmission from the WT; however, it increases the risks of losing raw historic data because of limited CMS memory size.

In order to acquire data that is directly comparable between each point and, importantly, to allow spectra to be recorded in apparently stationary conditions, a number of commercial CMSs, such as the SKF WindCon 3.0, can be configured to collect the vibration spectra within limited, pre-defined speed and power ranges. This is an important point to note when using traditional signal processing methods such as the FFT which require stationary signals within the analysis time window in order to obtain a clear result.

All the vibration-based CMSs surveyed have the capability to carry out some form of automatic diagnostic procedure. The majority of them are capable of producing alarms based either on the magnitude of spectral peaks, overall vibration levels or, in some cases, rates of oil debris particle generation. However,

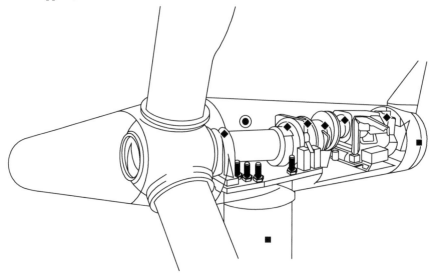

◆   Accelerometers.

■   Structural Vibration Monitoring (SVM) sensors conduct structural
    vibration analysis for monitoring and safety.

◉   The Data-Processing Front-end: acquires data and conducts
    analysis and alarm evaluations in real time for early detection
    of weakness or failure.

Figure 3.12 Typical sensor positions, adapted from (Gram & Juhl A/S, 2010)

results of automatic diagnosis must often be confirmed by vibration experts and component inspection. The level of confidence in these alarms is currently low but increases as monitoring engineers become more familiar with systems and turbines and as analysis techniques develop. Automatic diagnosis and prognosis are recent technologies which still require further investigation.

Only 6 out of the 27 vibration-based CMSs surveyed state that they are also able to monitor the level of debris particles in the gearbox coolant and lubrication oil system to enhance their CM capabilities. Modern oil debris counters take a proportion of the lubrication oil stream and detect and count both ferrous and non-ferrous particles of varying sizes. The counts can be fed as on-line data to the CMS. However, it should be noted that increasing measurement detail increases the cost of the on-line instrument.

Some recent commercially available CMSs are beginning to adapt to the WT environment and to be fully integrated into existing SCADA systems using standard protocols, examples are the GE Energy ADAPT.wind, the Romax Technology InSight Intelligent Diagnostic System and the Gram & Juhl TCM. Thanks to this integration, the analysis of the systems installed on the wind energy plant can also directly consider any other signals or variables of the entire controller network, for example current performance and operating condition, without requiring a doubling of the sensor system. The database, integrated into a single unified plant operations' view, allows a trend analysis of the condition of the machine. Recently patented condition-based turbine health monitoring systems, such as the Brüel & Kjaer VibroSuite, the Romax Technology InSight Intelligent Diagnostic System and the Gram & Juhl TCM, claim to feature diagnostic and prognostic software unifying fleet-wide CMS and SCADA, enabling the identification of both the source and cause of the fault and the application of prognostics to establish the remaining operational life of the component.

Only three on-line blade CMSs based on strain measurement using fibre optic transducers have been surveyed. These systems may be fitted to WT retrospectively. Compared with vibration monitoring techniques, these systems can be operated at low sampling rates as they are looking to observe changes in time domain. They are usually integrated in the WT control system but there are also some cases of integration, as an external input, into commercially available conventional vibration-based CMSs. As the blades continue to increase in cost and mass with the introduction of ever-larger wind machines, there is a great deal of concern about their reliability. It is believed that the development of reliable and effective blade monitoring systems will be a key enabler for future megawatt-scale turbines.

No commercial CMS is offered for electronics beyond oversight by the SCADA system. However, the reliability of both power electronics and electronic controls is of significant concern, especially for offshore installations where deterioration may be accelerated in the harsh environment by corrosion and erosion.

Cost justification

The cost of CMS based on vibration analysis, such as the SKF WindCon 3.0, is in the range of €15,000 to €20,000 per WT for software, transducers, cabling and installation (Nilsson and Bertling, 2007), more expensive than SCADA and with less coverage. The robustness with respect to failure detection/forecasting is not yet completely demonstrated and there has been considerable debate in the industry about the true value of CMS (Tavner, 2012). The cost justification of CMS for wind power has not been as clear as for traditional fossil-fired or nuclear power plants. To date, there have been few cost evidence publications to support the claims of the CMS industry. This is no doubt the result of data confidentiality within the industry.

A cost-benefit analysis by Walford (2006) shows that the lifetime savings derived from an early warning and the avoidance of the impending failures of the critical WT components would more than offset the lifetime cost of a CM system. Work by Besnard et al. (2010) shows results from a Life-Cycle-Cost model evaluated with probabilistic methods and sensitivity analysis to identify the economic benefit of using CMS in WTs. The results highlight that there is a high economic benefit of using CMS. The benefit is highly influenced by the reliability of the gearbox. Although the economics of deploying CM for a wind park is case dependent, some studies have shown the estimated return on assumed cost being better than 10:1 onshore, with total return on investment achieved in less than three years. These benefits will be even more dramatic if turbines are installed offshore where accessibility is a huge challenge (Sheng and Veers, 2011).

Yang et al. (2014) discuss gearbox failure costs and the cost advantages deriving from avoiding complete failure through successful WT CM for onshore and offshore individual 3 MW turbines and for an offshore WF, Scroby Sands (UK) comprising thirty 2 MW WTs. The work shows that, according to published failure rates, there is clearly a benefit in using WT CMS to eliminate gearbox failures. The figures for Scroby Sands offshore WF are particularly favourable and suggest that offshore WT CMS is essential for the avoidance of serious downtime and wasted offshore attendance. In addition, the authors notice that the inclusion of other major subassemblies, such as generator, blades and converter, would make a significant contribution to the WT CMS financial case, as would an extension of WF working life but all depends upon the reliability of the WT CMS and the ability of the operator to make use of its indications.

The investment in CM equipment for traditional power generation plants is normally covered by savings of costs from reduced unplanned production losses. For onshore WTs unplanned production losses are relatively low. Although for offshore WTs unplanned production losses are higher, the investment costs should be paid back by reduction of maintenance costs and reduced costs of increased damage (Hameed et al., 2009). According to Tavner (2012), WT CMS can only be justified if the system is capable of detecting a fault and giving early enough warning to avoid full subassembly replacement, which is the most costly aspect of failure, and if that CMS detection and warning can be acted upon by operators and WT OEMs to allow scheduling for repairs.

A practical example of CMS cost justification is given in Giordano and Stein (2011) which describes the real-life experience of an onshore 25-turbine WF operator using the 01dB-Metravib OneProd Wind CMS to successfully detect a broken tooth on the sun gear of the planetary stage of the gearbox, a generator bearing inner ring defect and a main bearing outer ring defect. This work demonstrated how the CMS investment cost was less than 1 per cent of the price of a current WT and the benefits coming from 17 per cent failure detection were sufficient to pay it back.

Current limitations and challenges of WT CMSs

Experience with CMSs in WFs to date is limited and shows that it is problematic to achieve reliable and cost-effective applications. The application of CM techniques has for decades been an integral part of asset management in other industries, and the technology has in recent years increasingly been adopted by the wind power industry. The general capabilities of the CM technology are therefore well known, but adaptation to the wind industry has proven challenging for successful and reliable diagnostics and prognostics as they are unmanned and remote power plants. There is still insufficient knowledge among WT maintenance staff of the potential of CMSs and inadequate experience of their application to common WT faults. The main differences which characterise WT operation compared with other industries are variable speed operation and the stochastic characteristics of aerodynamic load. This makes it difficult to use traditional frequency domain signal processing techniques, such as FFT, and to develop effective algorithms for early fault detection and diagnosis, due to the non-stationary signals involved. The majority of commercially available WT CMSs usually require experienced CM engineers able to successfully detect faults by comparing spectra at specific speeds and loads. Fault detection and diagnosis still require specialised knowledge of signal interpretation to investigate individual turbine behaviour, determine what analyses to perform and interpret the results with increased confidence. The lack of an ideal technique to analyse monitoring signals leads to frequent false alarms, which not only devaluate the WT CMS, but are also dangerous once a real fault occurs (Yang et al., 2009a).

One major limitation of the current commercially available CMSs is that very few operators make use of the alarm and the monitoring information available to manage their maintenance because of the volume and the complexity of the data. In particular, the frequent false alarms and costly specialist knowledge required for a manual interpretation of the complex vibration data, have discouraged WT operators from making wider use of CMSs. This happens despite the fact that these systems are fitted to the majority of the large WTs (> 1.5 MW) in Europe (Yang et al., 2014). Moreover, with the growth of the WT population, especially offshore, a manual examination and comparison of the CM data will be impractical unless a simplified monitoring process is introduced. The man-power costs of daily data analysis on an increasingly large WT population will be inappropriate to justify WT CM.

Most of the WT fault detection algorithms developed so far (Crabtree, 2011, Wiggelinkhuizen et al., 2008, Hameed et al., 2009, Hyers et al., 2006) require time consuming post-processing of monitored signals, which slows down the fault detection and diagnostic process, and still a certain degree of interpretation of the results. These algorithms are generally based on single signal analysis. Diagnosis can be difficult on the basis of a single signal alone while a multi-parameter approach, based on comparison of independent signals and able to recognise symptoms from different approaches, has shown an increased confidence in the practical applicability of these algorithms, potentially reducing the risk of false alarms (Feng et al., 2013).

Aspects of monitoring that particularly concern operators are the improvement of accuracy and reliability of diagnostic decisions, including level of severity evaluation, and the development of reliable and accurate prognostic techniques. It appears evident that cost effective and reliable CMSs would necessarily imply an increasing degree of data automation to deliver actionable recommendations in order to work effectively in the challenging offshore environment. The challenge is to achieve detection, diagnosis and prognosis as automatically as possible to reduce manpower and access costs. Current efforts in the wind CM industry are aimed at automating the data interpretation and improving the accuracy and the reliability of the diagnostic decisions, especially in the light of impending large-scale offshore WF generation.

The incorporation of refined efficient monitoring algorithms and techniques into existing CMSs could represent a way to increase their automation, enhance their capabilities, simplify and improve the accuracy and user confidence in alarm signals. In particular, automatic processing is required to separate multiple vibration components. These algorithms could reduce the quantity of information that WT operators must handle, providing improved detection and timely decision-making capabilities. Before being reported to the operators and asset managers, the raw data from remote CM stations would be processed and filtered by an on-line automatic acquisition system able to:

- detect incipient faults;
- diagnose their exact nature and give their location;
- ideally, provide a preliminary malfunction prognosis, through disciplined data management, sophisticated stochastic modelling and computational intelligence, in order to schedule a repair/replacement of the component before failure.

The operator could then choose to examine a particular WT in more detail, if required, so that the costly mobilisation of diagnosis specialists could be minimised only to serious, repairable WT faults. This tool will then allow for a simpler and faster analysis of the different signals transferred by the CMS since the expert will just receive information relevant to the signals showing defects.

In summary, the main advantages of automatic on-line CMSs with integrated fault detection algorithms are:

- appropriate management of WF data with a reduction of human workload, without losing accuracy in the phenomenon examination, and costs when handling high-rate monitoring information flow from disparate remote locations;
- the elimination of data post-processing;
- the application of a multi-parameter approach, able to compare at the same time independent monitoring signals, such as vibration signature, oil debris content and electric signature, providing reliable and timely automatic warnings/alarms of the incipient fault;
- the improvement of accuracy and reliability of the diagnostic decisions;
- the improvement of O&M strategy management according to the automatic prioritisation of fault severity set by reliable alarms.

Finally, as shown in the SUPERGEN Wind survey (Crabtree et al., 2014) provided in Appendix B, the majority of CMSs currently operate independently from the SCADA systems, therefore they do not use a lot of valuable information as operational parameters. It is then expected that the ultimate integration of autonomous CM and SCADA systems in the turbine controller will not only save costs but will also lead to more effective monitoring (Tavner, 2012).

Examples of wind turbine monitoring through CMSs

Several WT fault detection algorithms for the analysis of discretely sampled CM signals have been proposed in the research literature, including Djurović et al. (2012), Feng et al. (2013) and Zappalá et al. (2014). The fundamental idea of these algorithms is to reduce the computing demand for standard signal processing by tracking only significant features of WT non-stationary monitoring signals, rather than analysing wide frequency bandwidth signals. The implementation of the proposed algorithms can contribute to automate the fault detection and diagnosis process for the main WT components, improving the confidence in the alarms produced by the system and reducing uncertainty and risk in applying CM techniques directly to field data from operational WTs.

Djurović et al. (2012) investigate the influence of rotor electrical asymmetry on the stator line current and total instantaneous power spectra of WT wound rotor induction generators and doubly fed induction generators. The research is verified using experimental data measured on both Durham and Manchester test rigs and numerical predictions obtained from a time-stepped DFIG electromagnetic model. To give a clear indication of rotor electrical asymmetry in induction machines, a set of concise analytic expressions, describing fault frequency variation with operating speed, are defined and validated by measurement. Simulation and experimental results confirm that a convenient analysis and interpretation of identified fault frequencies in stator line current and power spectra lead to an effective rotor electrical fault detection and diagnostic procedure. In order to enable real-time fault frequency tracking a Fourier-based signal processing algorithm, the Iterative Localised Discrete Fourier Transform ($IDFT_{local}$) algorithm is proposed and applied to the experimental stator line current and total instantaneous power signals recorded from the Durham test rig driven under non-stationary, variable speed, wind-like conditions, as shown in Figure 3.13 and Figure 3.14, respectively. Although to a different extent for each frequency component, a step change in magnitude when the generator fault condition was present or has changed is clearly visible in both cases. These results show that the selected spectral components are valid fault indicators under variable load and speed conditions and also suggest that fault severity can be derived from their spectral component amplitude. This demonstrates the successful valid application of the proposed $IDFT_{local}$ algorithm for the analysis of fault-related speed-dependent frequencies within non-stationary signals such as those encountered on a WT.

Feng et al. (2013) present a case study of successful detection and diagnosis of gearbox bearing faults by a careful analysis of monitoring data recorded by the SKF WindCon CMS on 1.3 MW two-speed WTs. The results, presented in Figure 3.15, show how the gearbox bearing fault can be detected using multiple signals to improve detection confidence and using cumulative energy instead of the conventional signal time axis.

During period 'A', an early indication of the incipient fault is provided by the increasing independent signals, the particle generation rate and the vibration. The diagnosis is then confirmed by the simultaneous change in the two signals when entering period 'B'. During this period, an increasing rate of large particle generation is observed, suggesting significant material breakout from the bearing ferrous parts. The corresponding decrease in enveloped vibration suggests a deterioration in the vibration transmission path, due to bearing breakout material. This hypothesis was confirmed by a subsequent visual inspection, leading to the replacement of the gearbox ISS bearing. A significant vibration envelope magnitude reduction and a significant reduction in the 100–200 μm and 200–400 μm particle generation rate are observed in period 'C' following maintenance.

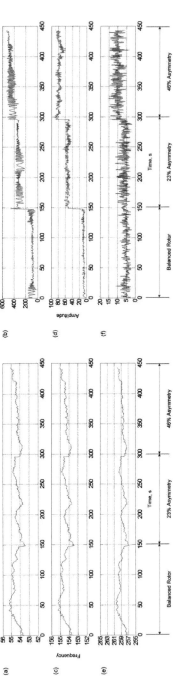

Figure 3.13 IDFT$_{local}$ analysis of stator line current frequencies of interest. Reproduced by permission of the Institution of Engineering & Technology (Djurović et al., 2012)

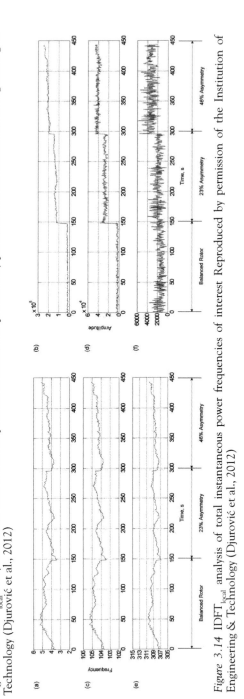

Figure 3.14 IDFT$_{local}$ analysis of total instantaneous power frequencies of interest Reproduced by permission of the Institution of Engineering & Technology (Djurović et al., 2012)

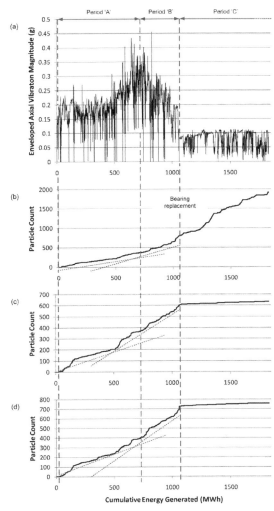

Figure 3.15 (a) Gearbox high-speed end enveloped axial vibration magnitude; and gearbox cumulative oil debris count for ferrous particles (b) 50–100 μm, (c) 100–200 μm, (d) 200–400 μm plotted against cumulative energy generated (Feng et al., 2013)

Zappalá et al. (2014) propose an advanced high-sensitivity algorithm, the gear Sideband Power Factor, SBPF$_{gear}$, algorithm, specifically designed to aid in the automatic on-line detection of gearbox faults when using field-fitted commercial WT CMSs. Uncertainty involved in analysing CM signals is reduced, and enhanced detection sensitivity is achieved, by identifying and collating characteristic fault frequencies in the vibration signal which can be tracked as the WT speed varies. The influence of the high-speed shaft gearbox pinion fault severity and the variable load operating conditions on the SBPF$_{gear}$ values is investigated by performing both constant and wind-like variable speed tests on

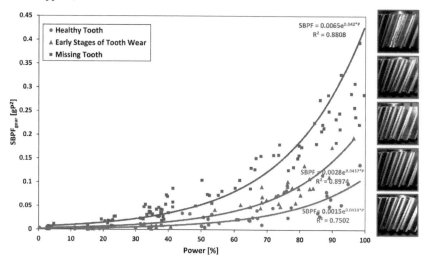

Figure 3.16 Influence of the high-speed shaft gearbox pinion fault severity and the variable load operating conditions on the $SBPF_{gear}$ values during the seeded-fault tests performed on the Durham WTCMTR, (adapted from Zappalá et al., 2014)

the Durham wind turbine condition monitoring test rig (WTCMTR), as shown in Figure 3.16. The results show that the proposed technique proves efficient and reliable for detecting both early and final stages of gear tooth damage, showing average detection sensitivities of 100 per cent and 320 per cent, respectively. The influence of the fault severity on the $SBPF_{gear}$ detection sensitivity values is evident; the more damaged the pinion the easier it is to discriminate the fault.

The performance of the experimentally defined technique is also successfully tested on signals from a full-size 750 kW WT gearbox that had experienced severe high-speed shaft gear set scuffing, with detection sensitivities of 1140 and 1251 per cent. Once implemented into WT CMSs, the proposed $SBPF_{gear}$ algorithm facilitates the monitoring analysis, reducing each FFT spectrum to only one parameter for each data acquisition. The generation of $SBPF_{gear}$ trends from the vibration spectra and the definition of magnitude thresholds for the fault severity levels could indicate to a WT operator when a maintenance action needs to be performed.

Appendix A: Survey of commercially available SCADA data analysis tools for wind turbine health monitoring

Bindi Chen, Donatella Zappalá, Christopher J. Crabtree and Peter J. Tavner

Introduction

A wind farm's existing Supervisory Control and Data Acquisition (SCADA) data stream is a valuable resource, which can be monitored by wind turbine (WT) manufacturers, operators and other experts to monitor and optimise

WT maintenance. In order to conduct an efficient SCADA data analysis, data analysis tools are required.

This survey discusses commercially available SCADA data analysis tools which are currently being applied in the WT industry. The main information is gathered from papers and internet with the help of the UK EPSRC Supergen Wind and EU FP7 ReliaWind Consortia. Information has been also collected from interaction with SCADA system manufacturers, WT manufacturers and product brochures at European Wind Energy Association (EWEA) Conference 2011, EWEA Conference 2012 and EWEA Conference 2013.

SCADA data

SCADA systems are a standard installation in large WTs and WFs – their data being collected from individual WT controllers. According to Zaher et al. (2009), the SCADA system assesses the status of the WT and its sub-systems using sensors fitted to the WT, such as anemometers, thermocouples and switches. The signals from these instruments are monitored and recorded at a low data rate, usually at 10 minute intervals. The SCADA data shows a snapshot of the WT operating condition at 10-minute intervals (Crabtree et al., 2014). Many large WTs are now also fitted with condition monitoring systems (CMSs), which monitor sensors associated with the rotating drive train, such as accelerometers, proximeters and oil particle counters. The CMS is normally separated from the SCADA system and collects data at much higher data rates, although some operators provide connections between the CMS and the WT controller to give the SCADA system some view of the CMS outputs.

A separate survey on commercially available WT CMSs has been conducted by Durham University and the UK SUPERGEN Wind Energy Technologies Consortium (Crabtree et al., 2014).

By analysing SCADA data, we are able to observe the relationship between different signals, and hence deduce the health of WT components. It would prove beneficial, from the perspective of utility companies, if SCADA data could be analysed and interpreted automatically to support the operators in identifying defects.

Commercially available SCADA data analysis tools

Table 3.A1 provides a summary of the available SCADA data analysis tools based on information collected from internet and from interaction with SCADA system manufacturers, WT manufacturers and product brochures at EWEA 2011, EWEA2012 and EWEA2013.

It should be noted that the table is as accurate as possible at the time of writing but may not be definitive. The products are arranged alphabetically by product name.

Table 3.A1 Table of commercially available SCADA Data Analysis Tools

1 **BaxEnergy WindPower Dashboard: BaxEnergy GmbH, Germany**
Description: Offers an extensive and comprehensive customisation for full integration of SCADA system applications and SCADA Software.
Main function: Real-time data acquisition and visualisation, alarm analytics and data reporting.

2 **CASCADA: Kenersys, Germany**
Description: Turbine control and visualisation in one system, applicable in a large number of control hardware solutions. Access to CASCADA possible from the control room or mobile devices. First 'open source solution' in the wind industry.
Main function: Smooth data integration via SQL database and individually customisable reports, alarms and availability calculations.

3 **CitectSCADA: Schneider Electric Pty. Ltd., Australia**
Description: CitectSCADA is a reliable, flexible and high performance system for any industrial automation monitoring and control application.
Main function: Graphical process visualization, superior alarm management, built-in reporting and powerful analysis tools.

4 **CONCERTO: AVL, Austria**
Description: A commercially available analysis and post-processing system capable of handling large quantities of data.
Main function: One tool for manifold applications, all data post-processing tasks within one tool, advanced data management.

5 **ENERCON SCADA system: ENERCON, USA**
Description: The Enercon SCADA System is used for data acquisition, remote monitoring, open-loop and closed-loop control for both individual WT and WFs. It enables the customer and Enercon Service to monitor the operating state and to analyse saved operating data.
Main function: Requesting status data, storing operating data, WF communication and loop control of the WF.

6 **Gamesa WindNet: Gamesa, Spain**
Description: The WindNet SCADA system in a WF is configured with a basic hardware and software platform based on Windows technology. The user interface is an easy to use SCADA application with specific options for optimal supervison and control of a WF including devices like WTs, meteorological masts and a substation.
Main function: Supervision and control of WTs, meteorological masts. Alarm and warning management, report generation and user management.

7 **Gateway System: Mita-Teknik, Denmark**
Description: Gateway is a PC-based software package, designed to collect, handle, analyse and illustrate the data from the controller with simple graphics and text. Operational parameters are recorded alongside with vibration signals/spectra and fully integrated into Gateway SCADA system.
Main function: Data collection and reporting, user-friendly operation, lifetime data, CMS algorithm toolbox for diagnostic analysis. Instant alarm notification. Approx 5000–8000 variables covering different production classes.

8 **GE – HMI/SCADA – iFIX 5.1: GE (General Electric Co.), USA**
Description: iFIX is a superior proven real-time information management and SCADA solution. It is open, flexible and scalable which includes impressive next-generation visualisations, a reliable control engine, a powerful built in historian and more.
Main function: Real-time data management and control, information analysis.

9 **GH SCADA: GL Garrad Hassan, Germany**
Description: GH SCADA has been designed by Garrad Hassan in collaboration with WT manufacturers, WF operators, developers and financiers to meet the needs of all those involved in WF operation, analysis and reporting.
Main function: Remote control of individual WT. Online data viewing. Reports and analysis.

10 **ICONICS for Renewable Energy: ICONICS Inc., USA**
Description: Provide portals for complete operations, including energy analytics, data histories and reports, geo SCADA with meteorological updates.
Main function: Portals for complete operations, data histories and reports, geo SCADA with meteorological updates.

11 **InduSoft Wind Power solutions: InduSoft, USA**
Description: InduSoft Web Sudio software consists of a powerful HMI/SCADA package that can monitor and adjust any operating set point in the controller or PLC.
Main function: WT monitoring, maintenance assistance and control room.

12 **INGESYS Wind IT: IngeTeam, Spain**
Description: INGESYS Wind IT makes it possible to completely integrate all the wind power plants into a single system. It provides advanced reporting services.
Main function: Advanced reporting, client/server architecture, standard protocols and formats.

13 **OneView SCADA system: SCADA International, Denmark**
Description: SCADA International is a subcontractor which provides specialised SCADA consultancy services both for wind power plant customers and WT manufactures, with services from commissioning of SCADA systems, project execution from sign of project to final handover and solutions to specific customer's requirements. It offers service contracts for SCADA systems beyond the manufacturer service and warranty.
Main function: Data collection, interface for viewing the data, advanced reporting tool, web access, eight standard reports (production, performance, availability, alarm, power curve, production graph, wind rose and raw data) and trend curves.

14 **PROZA NET: KONČAR, Croatia**
Description: Modular and multi-user system for control, monitoring and management of WTs. It enables supervisory control and execution of advanced functions (EMS, DMS, GMS) in real-time. Availability calculation, adjustable in accordance with owner/product contract, real-time trend view, power curve calculation, wind rose log, event log and history, alarm log and history and power distribution.
Main function: Remote monitoring of WT data such as temperatures, pitch and yaw, frequency converter and generator excitation, greasing and hydraulic brakes, vibration, cooling and heating, weather station and grid.

continued …

Table 3.A1 continued...

15 **reSCADA: Kinetic Automation Pty. Ltd., USA**
Description: Targeting and specialising in renewable energy industries. Will save time, effort and cost in developing HMI/SCADA.
Main function: Office 2007 GUI style, data visualisation, diary and mapping tools.

16 **SgurrTREND: SgurrEnergy, UK**
Description: SgurrEnergy provide a variety of wind monitoring solutions to evaluate the wind resource potential at a prospective WF site, offering a one-stop shop for all mast services from planning application, data collection and mast decommissioning to wind analysis services for energy yield prediction, project layout and design services.
Main function: Wind monitoring, processing and archiving the data, and reporting services.

17 **SIMAP: Molinos del Ebro, S.A., Spain**
Description: SIMAP is based on artificial intelligence techniques. It is able to create and dynamically adapt a maintenance calendar for the WT that is monitoring. The new and positive aspects of this predictive maintenance methodology have been tested in WTs.
Main function: Continuous collection of data, continuous processing information, failure risk forecasting and dynamical maintenance scheduling.

18 **VestasOnline Business SCADA: Vestas, Denmark**
Description: Power plant server continuously collects data from all the components and stores it in a central database. Data is processed in the server and fed to the standard operator interface through a local network, a wide area network or a dial-in connection. Communication network uses fibre-optic cables to allow communication between the server, clients, turbines, substations and the other units making up the wind power plant.
Main function: Monitoring and control of individual turbines, groups of turbines and the entire plant. Operator interface used to view current and historical power plant data. Advanced power plant regulation technology. Customisable with different types of interfaces.

19 **WindAccess: ALSTOM, Spain**
Description: Modular web-based tool. Remote access to WT data such as generated power, rotor rpm, electrical data, temperature of main components, mechanical sensors status, wind conditions and WT status. WF masts and substation equipment integration. Full support of Alstom WF experts. Performance optimisation of an individual WT or the whole WF in real time. Remotely collected data used to establish benchmarks and identify irregularities to allow timely intervention avoiding unplanned outages or secondary damage.
Main function: Alarm and warning management. Event and condition-based maintenance (ECBM) system combines planned maintenance with input from CMS and SCADA system. Appropriate schedule calculated taking into consideration weather forecasts, availability of spare parts and tools, as well as access and elevation logistics.

20 **Wind Asset Monitoring Solution: Matrikon, Canada**
Description: The solution bridges the gap between instrumentation and management systems, to enable operational excellence by retrieving and better managing data not readily accessible.
Main function: Monitor and manage all remote assets, leverage and integrate with SCADA and CMS.

21 **WindCapture: SCADA Solutions, Canada**
Description: WindCapture is a SCADA software package used for monitoring, controlling and data collection and reporting for WT generators. It was designed and tailored to the demands of manufacturers, operators, developers and maintenance managers of wind energy project and facilities.
Main function: Real-time data reporting with the highest degree of accuracy. Advanced GUI.

22 **WindHelm Portfolio Manager: GL Garrad Hassan, Germany**
Description: WindHelm Portfolio Manager provides a single platform for the monitoring, optimisation and control of any combination of operational turbines, farms and portfolios. It gives owners and operators uniform access to, and analysis of, their SCADA data, facilitating intelligent operational decisions.
Main function: Instant access to 'near real-time' data via a single user interface accessible from any web browser; broad range of summary and detailed operational reports; taxonomy from the RELIAWIND project included as standard; ability to send event alerts and status messages.

23 **Wind SCADA Pack for Renewable Energy: ICONICS, USA**
Description: Wind SCADA Pack allows users to create 2D and 3D GEO SCADA visualisation and reports with integrated real-time and historical geographical terrain maps, enabling a quick overview of multiple operations and plants located anywhere in the world. From one unified console wind park operators, engineers, and maintenance workers can monitor and control their entire operations.
Main function: Real-time turbine information, as wind speed, wind direction, power, blade position, temperature and vibration, are instantly data logged, visualized and analyzed. Users can receive weather information, power generated from turbines, react to alarm conditions and schedule maintenance.

24 **Wind Systems: SmartSignal, USA**
Description: SmartSignal analyses in real-time data and detects and notifies WFs of impending problems, allowing owners to focus on fixing problems early and efficiently.
Main function: Model maintenance, monitoring services, and predictive diagnostics.

25 **Wind Turbine In-Service: ABS Consulting, USA**
Description: The data gathered from inspections, vibration sensors and the SCADA system provide an overview of the WF and turbine conditions. The SCADA data allows monitoring WF assets in real-time, by also integrating historical and current data.
Main function: Regular diagnostics, dynamic performance reports, key performance indicators, fleet-wide analysis, forecasts and schedules and asset benchmarking assessment.

26 **Wind Turbine Prognostics and Health Management (WT-PHM) demonstration platform: American Center for Intelligent Maintenance Systems (IMS), USA**
Description: WT behavioural tools including routines for feature extraction, health assessment, and fault diagnostics. Results displayed in visualization tools for degradation assessment and monitoring. Multi-regime prognostics approach to handle the WT under various highly dynamic operating conditions
Main function: WT modelling for predictive maintenance. Platform implemented as a Watchdog Agent–based software platform. Hardware and instrumentation obtained using existing National Instruments tools.

A quick summary of Table 3.A1 shows that:

- six products are developed by WT manufacturers (2, 5, 6, 8, 18 and 19);
- five products are developed by renewable energy consultancies (9, 13, 16, 22 and 25);
- ten products are developed by industrial software companies (1, 3, 4, 10, 11, 15, 20, 21, 23 and 24);
- one product is developed by a WT operating company (17)
- three products are developed by an electrical equipment provider (7, 12, 14)
- one product is a demonstration platform developed by the American Center for Intelligent Maintenance Systems (IMS) (26)

Among these 26 products, Enercon SCADA System (5) and Gamesa WindNet (6) are examples of Wind Farm Cluster Management Systems (Wind on the Grid, 2009). Both provide a framework for data acquisition, remote monitoring, open/closed loop control, and data analysis for both individual WTs and WFs. Enercon SCADA System was launched in 1998 and is now used in conjunction with more than 11,000 WTs. Gamesa WindNet consists of a wide area network (WAN) system for WFs connected to an operational centre (Wind on the Grid, 2009).

GE – HMI/SCADA – iFIX 5.1 (8) was developed by General Electric Co. (GE), also a WT manufacturer. It is ideally suited for complex SCADA applications. The software also enables faster, better intelligent control and visibility of WF operations.

Kenersys CASCADA (2), VestasOnline Business SCADA (21) and Alstom WindAccess (22) are customisable SCADA systems also developed by WT manufacturers. CASCADA's open-source approach allows customers to service the turbines with external partners or enables them to do all the operation and maintenance work on their own. VestasOnline Business SCADA provides easy monitoring and control of individual turbines, groups of turbines and the entire plant. In WindAccess the remotely collected data can be used to establish benchmarks and identify irregularities allowing timely intervention to avoid unplanned outages or secondary damage.

GH SCADA (9), OneView SCADA system (13), SgurrTREND (16), WindHelm Portfolio Manager (22) and Wind Turbine In-Service (25) were developed by renewable energy consultancies in collaboration with WT manufacturers, WF operators, developers and financiers to meet the needs of all those involved in WF operation, analysis and reporting.

CONCERTO (4) is not specialised for SCADA data analysis. It is a generic data post-processing tool focusing on quick and intuitive signal analysis, validation, correlation and reporting for any kind of acquired data. Gray and Watson used it to perform analysis of WT SCADA data in (Gray and Watson, 2010).

SIMAP (17) is based on artificial intelligence techniques. The new and positive aspects of this predictive maintenance methodology have been tested on WTs. SIMAP has been applied to a WF owned by the Spanish wind energy company, Molinos del Ebro, S.A.

INGESYS Wind IT (12) was developed by IngeTeam, an electrical equipment provider. The system aims to integrate wind power plants into a single system and then optimise WF management. INGENSYS Wind IT also provides an advanced reporting service for power curve analysis, faults, alarms and customer reports.

Gateway System (7) was developed by another electrical equipment provider called Mita-Teknik. It is a PC-based software package, designed to collect, handle, analyse and illustrate the data from the WT controller with simple graphics and text. Operational parameters are recorded alongside with vibration signals/spectra and fully integrated into Gateway SCADA system.

PROZA NET (14) is the modular and multi-user SCADA system for control, monitoring and management of WTs presented by KONCAR, one of the biggest industrial companies in Croatia.

The other products – BaxEnergy WindPower Dashboard (1), CitectSCADA (3), ICONICS for Renewable Energy (10), InduSoft Wind Power (11), reSCADA (15), MATRIKON Wind Asset Monitoring Solution (20), WindCapture (21), Wind SCADA Pack for Renewable Energy (23) and Wind Systems (24) were developed by industrial software companies. These integrated SCADA systems aim to provide reliable, flexible and high performance applications for WT automation, monitoring and control.

Recent SCADA solutions, as (7), (25), can be adapted and fully integrated with commercial available conventional vibration-based CMSs using standard protocols. This unified plant operations' view allows a broad and complete analysis of the turbine's conditions by considering signals of the controller network as well as condition monitoring signals.

In some cases the SCADA product developer offers also service contracts for SCADA systems beyond the manufacturer service and warranty. They usually include hardware audit, system specific maintenance plan, monthly check of the SCADA system and 24/7 online support. Examples are ABS Wind Turbine In-Service (25), SCADA International OneView SCADA system (13) and several others.

The Wind Turbine Prognostics and Health Management (26) demonstration platform has been developed by the American Center for Intelligent Maintenance Systems (IMS) for asset health information by intelligent interpretation of SCADA measurements. It features WT modelling for predictive maintenance and a multi-regime prognostics approach to handle the WT under various highly dynamic operating conditions.

Conclusions

From this survey we conclude that:

- there is a wide variety of commercial SCADA systems and SCADA data analysis tools available to the wind industry;
- most of the commercially available SCADA data analysis tools are able to analyse real-time data;
- the performance analysis techniques used in the SCADA data analysis tools vary from tailored statistical methods to the use of artificial intelligence;
- successful SCADA data analysis tools provide cluster management for WFs; they provide a framework for data acquisition, alarm management, reporting and analysis, production forecasting and meteorological updates;
- some built-in diagnostics techniques are able to diagnose the component failure of WT;
- some SCADA data analysis tools are beginning to be fully integrated by operators with commercially available CMSs;
- recently proposed SCADA systems feature WT modelling, with diagnostic and prognostic models for WF predictive maintenance.

Finally, it should be noted that the development of SCADA data analysis tools is aimed to provide a reliable, flexible and high performance for WT automation monitoring and control. The industry is already noting the importance of operational parameters such as load and speed and so techniques may begin to adapt further to the WT environment leading to more reliable WT diagnostics solution (Crabtree et al., 2014).

Appendix B: Survey of commercially available condition monitoring systems for wind turbines

Christopher J. Crabtree, Donatella Zappalá and Peter J. Tavner

Introduction

As wind energy assumes greater importance in remote and offshore locations, affective and reliable condition monitoring (CM) techniques are required. Conventional CM methods used in the power generation industry have been adapted by a number of industrial companies and have been applied to wind turbines (WT) commercially.

This survey discusses commercially available condition monitoring systems (CMS) which are currently being applied in the WT industry. Information has been gathered over several years from conferences and websites and includes information available from product brochures, technical documents and discussion with company representatives. The research was carried out as part of:

- Theme X of the SUPERGEN Wind Energy Technologies Consortium, Phase 1, whose objective was to devise a comprehensive CMS for practical application on WTs.

- Theme 2 of the SUPERGEN Wind Energy Technologies Consortium, Phase 2, whose objective, built on the work in SUPERGEN Wind 1, was to develop turbine monitoring targeted at improving the reliability and availability of offshore WFs.
- Theme 4.3 of the SUPERGEN Wind Energy Technologies Consortium, Phase 2, whose objective was to develop fault identification methodologies for electrical and mechanical drivetrain systems.

The report also identifies some of the advantages and disadvantages of existing commercial CMSs alongside discussion of access, cost, connectivity and commercial issues surrounding the application of WT CMSs.

Reliability of wind turbines

Quantitative studies of WT reliability have recently been carried out based on publicly available data (Tavner et al., 2007; Spinato et al., 2009; Ribrant and Bertling, 2007). These studies have shown WT gearboxes to be a mature technology with constant of slightly deteriorating reliability with time. This would suggest that WT gearboxes are not an issue, however surveys by WMEP and LWK (Faulstich et al., 2008) have shown that gearboxes exhibit the highest downtime per failure among onshore sub-assemblies. This is shown graphically in Figure 3.B1 where we clearly see consistently low gearbox failure rate between two surveys with high downtime per failure. Similar results have also been shown for the Egmond aan Zee wind farm (NoordzeeWind, 2007-2009) where gearbox failure rate is not high but the downtime and resulting costs are. The poor early reliabilities for gearbox and

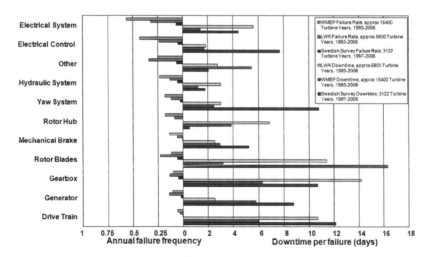

Figure 3.B1 Wind turbine sub-assembly failure rate and downtime per failure from three surveys including over 24000 turbine years of data as published in (Ribrant and Bertling, 2007; Tavner et al., 2010)

drive train reliability components have led to an emphasis in WT CMSs on drive train components and therefore on vibration analysis.

The high downtime for gearboxes derives from complex repair procedures. Offshore WT maintenance can be a particular problem as this involves specialist equipment such as support vessels and cranes but has the additional issue of potentially unfavourable weather and wave conditions. The EU-funded project ReliaWind has developed a systematic and consistent process to deal with detailed commercial data collected from operational WFs. This includes the analysis of 10-minute average SCADA data as discussed earlier, automated fault logs, and operation and maintenance reports. The research aimed to identify and understand WT gearbox failure mechanisms in greater detail (Wilkinson et al., 2010). However, more recent information on WT reliability and downtime, especially when considering offshore operations suggests that the target for WT CMSs should be widened from the drive train towards WT electrical and control systems (Wilkinson et al., 2011).

As a result of low early reliability, particularly in large WTs, and as a result of the move offshore, interest in CMSs has increased. This is being driven forward by the insurer Germanischer Lloyd who published guidelines for the certification of CMSs (Germanischer Lloyd, 2007) and certification of WTs both onshore (Germanischer Lloyd, 2003) and offshore (Germanischer Lloyd, 2005).

Monitoring of wind turbines

WTs are monitored for a variety of reasons. There are a number of different classes into which monitoring systems could be placed. Figure 3.B2 shows the general layout and interaction of these various classes.

First, we have SCADA systems. Initially these systems provided measurements for WT energy production and to confirm that the WT was operational through 5–10 minute averaged values transmitted to a central database. However, SCADA systems can also provide warning of impending malfunctions in the WT drive train. According to Zaher et al. (2009) the 10-minute averaged signals usually monitored in modern SCADA systems include:

- active power output (and standard deviation over 10-minute interval);
- anemometer-measured wind speed (and standard deviation over 10-minute interval);
- gearbox bearing temperature;
- gearbox lubrication oil temperature;
- generator winding temperature;
- power factor;
- reactive power;
- phase currents;
- nacelle temperature (1-hour average).

This SCADA configuration is designed to show the operating condition of a WT but not necessarily give an indication of the health of a WT. However, more

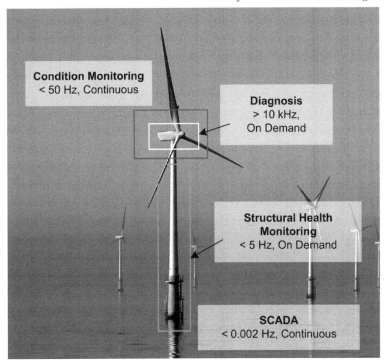

Condition Monitoring
< 50 Hz, Continuous

Diagnosis
> 10 kHz,
On Demand

Structural Health Monitoring
< 5 Hz, On Demand

SCADA
< 0.002 Hz, Continuous

Figure 3.B2   Structural health and condition monitoring of a wind turbine

up to date SCADA systems include additional alarm settings based not only on temperature but also on vibration transducers. Often we find several vibration transducers fitted to the WT gearbox, generator bearings and the turbine main bearing. The resultant alarms are based on the level of vibration being observed over the 10-minute average period. Research has being carried out into the CM of WTs through SCADA analysis in the EU project ReliaWind (ReliaWind, 2011). The research consortium consisted of a number of university partners alongside industrial consultants and WT manufacturers.

Second, there is the area of structural health monitoring (SHM). These systems aim to determine the integrity of the WT tower and foundations. SHM is generally carried out using low sampling frequencies below 5Hz.

While SCADA and SHM monitoring are key areas for WT monitoring, this survey will concentrate on the remaining two classes of CM and diagnosis systems.

Monitoring of the drive train is often considered to be the most effective through the interaction of these two areas. CM itself may be considered as a method for determining whether a WT is operating correctly or whether a fault is present or developing. A WT operator's main interest is likely to be in obtaining reliable alarms based on CM information which can enable them to take confident action with regard to shutting down for maintenance. The operator need not know the exact nature of the fault but would be alerted to the severity of the issue by the alarm signal. Reliable CM alarms will be essential for any operator with a large number of WTs under its ownership. On this basis, CM signals need not be collected on a high frequency basis as this will reduce bandwidth for transmission and space required for storage of data.

Once a fault has been detected through a reliable alarm signal from the CMS, a diagnosis system could be activated either automatically or by a monitoring engineer to determine the exact nature and location of the fault. For diagnosis systems, data recorded at a high sampling frequency is required for analysis; however, this should only be collected on an intermittent basis. The operational time of the system should be configured to provide enough data for detailed analysis but not to flood the monitoring system or data transmission network with excess information.

Finally, Figure 3.B3 gives an indication of three sections of a WT that may require monitoring based on reliability data such as that in Figure 3.B1 (Ribrant and Bertling, 2007; Tavner et al., 2010). While each of the three areas are shown as separate entities, CM must blur the boundaries between them in order to provide clear alarms and, subsequently, diagnostic information.

Many of the CMSs included in this survey are a combination of CMSs and diagnostic systems due to the high level of interaction that can exist between the two types of system.

Commercially available condition monitoring systems

Table 3.B1 provides a summary of a number of widely available and popular CMSs for WTs. The information in this table has been collected from interaction with CMS manufacturers, WT manufacturers and product brochures over a long period of time and is up to date as of the time of writing. However, since some information has been acquired through discussion with sales and product representatives and not from published brochures, it should be noted that the table may not be fully definitive and is only as accurate as the given information. The systems in Table 3.B1 are arranged alphabetically by product name.

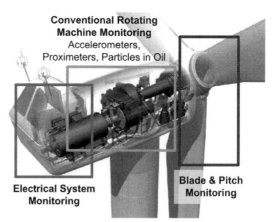

Figure 3.B3 General layout of three areas for condition monitoring and diagnosis within the nacelle of a WT

Table 3.B.1 Table of commercially available condition monitoring systems. (based on available literature and contact with industry including EWEA Conference 2008, 2009, 2010, 2011, 2012, 2013, 2014)

1 **ADAPT.wind: GE Energy, USA**
Description: Up to 150 static variables monitored and trended per WT. Planetary cumulative impulse detection algorithm to detect debris particles through the gearbox planetary stage. Dynamic energy index algorithm to spread the variation over five bands of operation for spectral energy calculations and earlier fault detection. Sideband energy ratio algorithm to aid in the detection of gear tooth damage. Alarm, diagnostic, analytic and reporting capabilities facilitate maintenance with actionable recommendations. Possible integration with SCADA system.
Main components monitored: Main bearing, gearbox, generator
Monitoring technology: Vibration (accelerometer); oil debris particle counter
Analysis method(s): FFT frequency domain analysis; time domain analysis
Data or sampling frequency: N/A

2 **APPA System: OrtoSense, Denmark**
Description: Oscillation technology based on interference analysis that replicates the human ear's ability to perceive sound.
Main components monitored: Main bearing, gearbox, generator
Monitoring technology: Vibration
Analysis method(s): Auditory perceptual pulse analysis (APPA)
Data or sampling frequency: N/A

3 **Ascent; Commtest, New Zealand**
Description: System available in three complexity levels. Level 3 includes frequency band alarms, machine template creation and statistical alarming.
Main components monitored: Main shaft, gearbox, generator
Monitoring technology: Vibration (accelerometer)
Analysis method(s): FFT frequency domain analysis, envelope analysis, time domain analysis
Data or sampling frequency: N/A

4 **Brüel & Kjaer Vibro: Brüel & Kjaer(Vestas), Denmark**
Description: Local data acquisition units, alarm management and review by the Condition Monitoring Centre analysts, reports on actionable information to customers. Vibration and process data automatically monitored at fixed intervals and remotely sent to the diagnostic server. Monitoring to specific power loadings and filtering out irrelevant alarms. Time waveform automatically stored before and after user-defined event for advanced vibration post-analysis. Severity classes, each related to an estimated lead-time. Severity is first estimated automatically by the Alarm Manager, followed by the diagnostic expert final assessment.
Main components monitored: Main bearing, coupling, gearbox, generator, nacelle, support structure, nacelle temperature, noise in the nacelle
Monitoring technology: Vibration; temperature sensor; acoustic
Analysis method(s): Time domain; FFT frequency analysis
Data or sampling frequency: Variable up to 40kHz.; 25.6kHz.

5 **Brüel & Kjaer VibroSuite: Brüel & Kjaer (Vestas),Denmark**
Description: Stand-alone software packages, completely client-owned, enable end-users and operators to host, process and analyse the data in-house. AlarmManager processes and simplifies data, provides developing faults automated evaluation, severity level and lead-time to failure; AlamTracker for quick access to live

continued ...

Table 3.B.1 continued...

alarms with alarm history and high-end functionality; WTG.Analyser diagnostic tool interprets signals and identifies root causes; EventMaster facilitates time and event based diagnostic data acquisition.
Main components monitored: Main bearing, coupling, gearbox, generator, nacelle, support structure, nacelle temperature, noise in the nacelle
Monitoring technology: Vibration; temperature sensor; acoustic
Analysis method(s): Time domain; FFT frequency analysis
Data or sampling frequency: Variable up to 40kHz.; 25.6kHz.

6 **CMS: Nordex, Germany**
Description: Start-up period acquires vibration 'fingerprint' components. Actual values automatically compared by frequency, envelope and order analysis, with the reference values stored in the system. Some Nordex turbines also use the Moog Insensys fibre optic measurement system.
Main components monitored: Main bearing, gearbox, generator
Monitoring technology: Vibration (accelerometer)
Analysis method(s): Time domain based on initial 'fingerprint'
Data or sampling frequency: N/A

7 **Condition Based Maintenance System (CBM): GE (Bently Nevada), USA**
Description: This is built upon the Bently Nevada ADAPT.wind technology and System 1. Basis on System 1 gives monitoring and diagnostics of drive train parameters such as vibration and temperature. Correlate machine information with operational information such as machine speed, electrical load, and wind speed. Alarms are sent via the SCADA network.
Main components monitored: Main bearing, gearbox, generator, nacelle, optional bearing and oil temperature
Monitoring technology: Vibration (accelerometer)
Analysis method(s): FFT frequency domain analysis; acceleration enveloping
Data or sampling frequency: N/A

8 **Condition Based Monitoring System: Bachmann electronic GmbH, Austria**
Description: Up to nine piezoelectric acceleration sensors per module. Basic vibration analysis with seven sensors. PRÜFTECHNIK solid borne sound sensors for low frequency diagnostics of slowly rotating bearings on the WT LSS. Three channels for the ±10V standard signal per module. Audibile sensor signals to assess the spectra acoustically. Fully integration in automation control system to link the measured values to operating parameters and increase diagnostic reliability. Traffic light system indicates if predefined thresholds are exceeded. Data analyzed by experienced diagnostics specialists using extensive tools, such as envelope and amplitude spectra, or frequency-based characteristic values. Integrated database enables data trend analysis.
Main components monitored: Main drive train components, generator
Monitoring technology: Vibration (accelerometer); acoustic
Analysis method(s): Time domain; FFT frequency analysis
Data or sampling frequency: 24-bit res; 190 kHz sample rate per channel; 0.33 Hz (solid borne sound sensors)

9 **Condition Diagnostics System: Winergy, Denmark**
Description: Up to six inputs per module. Advanced signal processing of vibration levels, load and oil to give automated machinery health diagnostics, forecasts and recommendations for corrective action. Automatic fault identification is provided. Relevant information provided in an automated format to the operations and maintenance centre, without any experts being involved. Information delivered to the appropriate parties in real time. Pitch, controller, yaw and inverter monitoring can also be included.

Main components monitored: Main shaft, gearbox, generator
Monitoring technology: Vibration (accelerometer); oil debris particle counter
Analysis method(s): Time domain; FFT frequency domain analysis
Data or sampling frequency: 96kHz per channel

10 **Condition Management System (CMaS): Moventas, Finland**
Description: Compact remote system measuring temperature, vibration, load, pressure, speed, oil aging and oil particle count. 16 analogue channels can be extended with adapter. Performance monitoring, anticipate possible upcoming failures by providing timely updates and alerting maintenance crews. Data stored, analysed and reported to remote server via standard TCP/IP protocol. Mobile interface available. Moventas Remote Centre provides proactive gear expertise with specialists available for all customers on-call 24 hours a day, seven days a week.
Main components monitored: Gearbox, main bearing, generator, rotor, turbine controller
Monitoring technology: Temperature; vibration; load; pressure; RPM; oil condition/particles
Analysis method(s): Time domain (possible FFT)
Data or sampling frequency: N/A

11 **Distributed Condition Monitoring System: National Instruments, USA**
Description: Up to 32 channels; default configuration: 16 accelerometer/ microphone, four proximity probe and eight tachometer input channels. Also provided mixed-measurement capability for strain, temperature, acoustics, voltage, current and electrical power. Oil particulate counts and fiber-optic sensing can also be added to the system. Possible integration into SCADA systems.
Main components monitored: Main bearing, gearbox, generator
Monitoring technology: Vibration; acoustic
Analysis method(s): Spectral analysis; level measurements; order analysis; waterfall plots; order tracking; shaft centre-line measurements; Bode plots
Data or sampling frequency: 24-bit res; 23.04 kHz of bandwidth with antialiasing filters per accelerometer/microphone channel

12 **HAICMON: Hainzl, Austria**
Description: The monitoring unit (CMU), mounted in the nacelle, performs the data acquisition, analysis and local intermediate storage. Up to 32 vibration inputs, 8 digital and optional 16 analog inputs. Web interface for configuration and visualization purpose. Automatic data analysis directly on the CMU with automatic alarming features. CMU can operate in standalone mode or in connection with the superior HAICMON Analysis Center which features more computing power or database access functionality for advanced trend analysis. It also allows comparing different plants among each other and provides a reporting module.
Main components monitored: Rotor bearing, gearbox, generator
Monitoring technology: Vibration; load; rotation speed; oil temperature
Analysis method(s): Time domain; FFT frequency analysis; envelope analysis; cepstrum analysis
Data or sampling frequency: Variable up to 40kHz.

13 **InSight intelligent Diagnostic System (iDS): Romax Technology Ltd, UK**
Description: Diagnostic and prognostic software unifying fleet wide CMS and SCADA data. Suite of predictive maintenance technologies and services comprising: Inspection and Analysis; iDS that integrates vibration, SCADA and maintenance record data. This hardware independent software platform harmonises data from multiple manufacturers CMSs. Intuitive user interface and

continued …

Table 3.B.1 continued...

advanced diagnostic rules. iDS Manger provides managers with a clear dashboard displaying WTs' condition and notifying of alarm events via email. InSight Expert is a diagnostic platform aimed at vibration analysis experts which enables the identification of both fault source and cause and the application of prognostics.
Main components monitored: Main drive train components
Monitoring technology: Vibration; temperature; oil debris particle counter
Analysis method(s): Time domain; FFT frequency analysis
Data or sampling frequency: N/A

14 **OMNITREND: Prüftechnik, Germany**
Description: WebReport creates customizable reports for analysing machine conditions, color-coded alarm classes in the status report identify machine problems at a glance; Online View, visualizes measurement data from online systems and machine conditions in real time; network capable multiuser PC software OMNITREND saves measurement data in a database, arranges routes for data collection and visualizes the results in easy-to-read diagrams. Practical tools support data evaluation and documentation. Data exchange between OMNITREND and a computerised maintenance management system allows exchanging measurement data, sending status messages and using master data from ERP systems.
Main components monitored: Main drive train components
Monitoring technology: Vibration
Analysis method(s): FFT frequency domain analysis
Data or sampling frequency: N/A

15 **OneProd Wind System: ACOEM (01dB-Metravib), France**
Description: 8 to 32 channels; operating conditions trigger data acquisitions. Repetitive and abnormal shock warnings enable detection of failure modes; built-in diagnostic tool. Optional additional sensors for shaft displacement and permanent oil quality monitoring; structure low frequency sensors; current&voltage sensors. Graphics module for vibration analysis including many representation modes as trend, simple or concatenated spectrum, waterfall spectrum, time signal, circular view, and orbit; advanced cursors and post-processing are available. SUPERVISION web application provides information on alarm status together with expert diagnoses and recommendations.
Main components monitored: Main bearing on LSS, bearing on gearbox LSS, bearing on intermediate gearbox shaft, on gearbox high-speed shaft, on generator; oil debris, structure, shaft displacement, electrical signals
Monitoring technology: Vibration; acoustic; electrical signals; thermography; oil debris particle counter
Analysis method(s): Time domain; FFT frequency analysis; electrical signature analysis
Data or sampling frequency: N/A

16 **SIPLUS CMS4000: Siemens, Germany**
Description: Acquisition and evaluation of analog and binary signals in individual WTs or complex WFs. Data acquisitions and pre-processing performed with interface nodes that enable the recording of highly dynamic processes. Software X tools for analysis, diagnostics, visualisation and archiving for wind power plants. Data displayed via traffic lights, integrated message systems and spectrum-view in CMS X-Tools. Library of standard function blocks for FFT, envelope curve analysis, input filters, mathematical and communication functions and graphical creation of diagnostic models. Analysis blocks can be interconnected graphically to resolve specific measuring and diagnostic tasks. Modular, scalable system can be integrated into existing WTs and new ones.

Main components monitored: Bearings , gearbox, tower
Monitoring technology: Vibration
Analysis method(s): FFT frequency domain analysis; envelope curve analysis;
fingerprint comparison; trend analysis; input filters
Data or sampling frequency: Exceeding 40kHz

17 **SMP-8C: Gamesa Eolica, Spain**
Description: Continuous online vibration measurement of main shaft, gearbox
and generator. Comparison of spectra trends. Warnings and alarm transmission
connected to WF management system.
Main components monitored: Main shaft, gearbox, generator
Monitoring technology: Vibration
Analysis method(s): FFT frequency domain
Data or sampling frequency: N/A

18 **System 1: Bently Nevada (GE), USA**
Description: Monitoring and diagnostics of drive train parameters such as
vibration and temperature. Correlate machine information with operational
information such as machine speed, electrical load, and wind speed.
Main components monitored: Main bearing, gearbox, generator, nacelle. Optional
bearing and oil temperature
Monitoring technology: Vibration (accelerometer)
Analysis method(s): FFT frequency domain; acceleration enveloping
Data or sampling frequency: N/A

19 **TCM (Turbine Condition Monitoring) Enterprise V6 Solution with SCADA
integration: Gram & Juhl A/S, Denmark**
Description: Advanced signal analysis and process signals combined with
automation rules and algorithms for generating references and alarms. M-System
hardware features up to 24 synchronous channels, interface for structural
vibration monitoring and RPM sensors, extern process parameters and analog
outputs. TCM Site Server stores data and does post data processing (data
mining) and alarm handling. TCM Ocular Modeller models drive train and
sensor configuration to relate measurement data to the kinematics of the turbine.
Control room with web based operator interface. TCM Enterprise allows
centralised remote monitoring on a global scale. Optional Structural Vibration
Monitoring sensor to measure low frequency signals associated with tower sway,
rotor imbalance and machine over speed. Integration with SCADA through
OPC UA.
Main components monitored: Tower, blades, shaft, nacelle, main bearing, gearbox,
generator
Monitoring technology: Vibration (accelerometer); sound analysis; strain analysis;
process signals analysis
Analysis method(s): FFT and wavelet frequency domain analysis; envelope, time
and frequency domain (analytic/Hilbert) analysis; cepstrum, kurtosis, spectral
kurtosis, skewness; RMS analysis; order tracking analysis
Data or sampling frequency: 40.960/81.920 kHz

20 **TurbinePhD (Predictive Health Monitoring): NRG Systems, USA**
Description: Automatically integrates multiple condition indicators into a single
readily understandable health indicator for each turbine's drive train component
delivering future health predictions. Actionable indicators and data supporting
the diagnosis accessible via Internet. Optimise turbine maintenance schedules
by predicting when components in the turbine's drive train are likely to fail
and scheduling repairs at the most cost-effective time. Advanced diagnostic
algorithms from the aerospace industry accounting for varying speed and torque

continued …

Table 3.B.1 continued...

conditions. Residual, energy operator, narrow band and modulation analysis tools for gear analysis; spectrum and envelope for bearing analysis; synchronous average for shaft analysis.
Main components monitored: Main shaft, gearbox, main bearing
Monitoring technology: Vibration (accelerometer)
Analysis method(s): FFT frequency domain analysis
Data or sampling frequency: 0.78 to 100 Hz @24 bits (high speed vibration sensor); 8 to 500 hz (low speed vibration sensor)

21 **VIBstudio WIND: EC Systems, KAStrion project, Poland**
Description: Integrated embedded system for data acquisition; real-time verification through algorithms for automatic signal validation, to avoid generating false alarms, and advanced signal processing. Up to 24 vibration channels, four analogue channels, two digital inputs and three digital outputs. Automated generation of analyses and thresholds; individually tuned, automated configuration of machine operational state; intelligent data selection and storage; tolerance for loss of connectivity. VIBmonitor Astrion module for automatic vibration analysis. VIBmonitor SMESA module for generator fault detection by electrical signature analysis.
Main components monitored: Bearings, shaft, gearbox, generator
Monitoring technology: Vibration (accelerometer)
Analysis method(s): Wideband analyses: PP, RMS, VRMS, crest, kurtosis; narrowband analyses: energy in the band, order spectrum and envelope spectrum
Data or sampling frequency: Vibration channels: variable up to 100 kHz; process variable channels: up to 1 kHz

22 **Wind AnalytiX: ICONICS, USA**
Description: This software solution uses fault detection and diagnostics technology which identifies equipment and energy inefficiencies and provides possible causes that help in predicting plant operations, resulting in reduced downtime and costs related to diagnostic and repair.
Main components monitored: Main WT components
Monitoring technology: Vibration (accelerometer)
Analysis method(s): Unknown
Data or sampling frequency: N/A

23 **WindCon 3.0: SKF, Sweden**
Description: Monitoring solution including sensors, data export, analysis and lubrication. Turbine health monitoring through vibration sensors and access to the turbine control system by means of the SKF WindCon software. WindCon 3.0 collects, analyses and compiles operating data that can be configured to suit management, operators or maintenance engineers. The system can be stand alone or linked together using SKF's WebCon, the web solution for data hosting and remote monitoring. WindCon can also be linked to the turbine lubrication system and fully integrated with the WindLub system for automated condition based lubrication and monitoirng of the lubrication pump.
Main components monitored: Blade, main bearings, shaft, gearbox, generator, tower, generator electrical
Monitoring technology: Vibration (accelerometer, proximity probe); oil debris particle counter
Analysis method(s): FFT frequency domain analysis; envelope analysis; time domain analysis
Data or sampling frequency: Analogue: DC to 40kHz (variable, channel dependent); digital: 0.1 Hz–20kHz

24 **Wind Turbine In-Service: ABS Consulting, USA**
Description: Data gathered from inspections, vibration sensors and SCADA system. Ekho for WIND software features regular diagnostics, dynamic performance reports, key performance indicators, fleet-wide analysis, forecasts/schedules, and asset benchmarking. It generates alarms and notifications or triggers work orders for inspections or repairs.
Main components monitored: Main bearing, gearbox, generator, gearbox and gear oil, rotor blades and coatings
Monitoring technology: Vibration; inspections
Analysis method(s): FFT frequency domain analysis; time domain analysis
Data or sampling frequency: N/A

25 **WinTControl: Flender Service GmbH, Germany**
Description: Vibration measurements are taken when load and speed triggers are realised. Time and frequency domain analysis are possible.
Main components monitored: Main bearing, gearbox, generator.
Monitoring technology: Vibration (accelerometer)
Analysis method(s): FFT frequency domain; time domain analysis
Data or sampling frequency: 32.5kHz

26 **WiPro: FAG Industrial Services GmbH, Germany**
Description: Measurement of vibration and other parameters given appropriate sensors. Time and frequency domain analysis carried out during alarm situations. Allows speed-dependent frequency band tracking and speed-variable alarm level.
Main components monitored: Main bearing, shaft, gearbox, generator, temperature (adaptable inputs)
Monitoring technology: Vibration (accelerometer)
FFT frequency domain; time domain analysis
Data or sampling frequency: Variable up to 50kHz

27 **WP4086: Mita-Teknik, Denmark**
Description: Up to eight accelerometers for real-time frequency and time domain analysis. Warnings/Alarms set for both time and frequency domains based on predefined statistical/thresholds-based vibration limits. Operational parameters recorded alongside with vibration signals/spectra and full integration into Gateway SCADA system. Algorithm Toolbox for diagnostic analysis. Approx 5000–8000 variables covering different production classes.
Main components monitored: Main bearing, gearbox, generator
Monitoring technology: Vibration (accelerometer)
Analysis method(s): FFT amplitude spectra; FFT envelope spectra; time domain magnitude; comb filtering, whitening, Kurtogram analysis
Data or sampling frequency: 12-bit chan res; variable up to 10kHz

28 **CMSWind: CMSWind project, European Research & Development Project**
Description: (Still in development phase) Advanced condition monitoring system which utilises motor current signature analysis, operational modal analysis and acoustic emission techniques to monitor the condition of the generator, the gearbox and rotary components, respectively. All systems are tied together through SCADA to provide supervisory control, data logging and analysis. Wireless sensors for rotating components monitoring using high performance powering and energy harvesting technologies.
Main components monitored: Gearbox (including main bearing, yaw system, hub), generator

continued …

Table 3.B.1 continued...

Monitoring technology: Electrical signals; operational parameters, acoustic
Analysis method(s): Motor current signature analysis; operational modal analysis ;
acoustic emission techniques
Data or sampling frequency: N/A

29 **HYDACLab: HYDAC Filtertechnik GmbH, Germany**
Description: Permanent monitoring system to monitor particles (including air
bubbles) in hydraulic and lube oil systems.
Main components monitored: Lubrication oil and cooling fluid quality
Monitoring technology: Oil debris particle counter
Analysis method(s): N/A
Data or sampling frequency: N/A

30 **Oil Contamination Monitor (OCM 30X): C.C. Jensen, Denmark**
Description: Early warning for gearbox breakdown by measuring wear generation.
Especially designed for high viscous oils, such as gear oils, and equipped with
an air removal device to enable correct measurements. Very stable flow over
a large viscosity range allows sensor accurate readings. Different options for
communication with the SCADA system and wen based trends.
Main components monitored: Lubrication oil quality and cleanliness
Monitoring technology: Oil debris particle counter; oil cleanliness sensor
Analysis method(s): N/A
Data or sampling frequency: N/A

31 **PCM200: Pall Industrial Manufacturing, (Pall Europe Ltd, UK)**
Description: Fluid cleanliness monitor reports test data in real-time so ongoing
assessments can be made. Can be permanently installed or portable.
Main components monitored: Lubrication oil cleanliness
Monitoring technology: Oil cleanliness sensor
Analysis method(s): N/A
Data or sampling frequency: N/A

32 **TechAlert 10/TechAlert 20: MACOM, UK**
Description: TechAlert 10 is an inductive sensor to count and size ferrous and
non-ferrous debris in circulating oil systems. TechAlert 20 is a magnetic sensor
to count ferrous particles.
Main components monitored: Lubrication oil quality
Monitoring technology: Inductive or magnetic oil debris particle counter
Analysis method(s): N/A
Data or sampling frequency: N/A

33 **BLADEcontrol: IGUS ITS GmbH, Germany**
Description: Accelerometers are bonded directly to the blades and a hub
measurement unit transfers data wirelessly to the nacelle. Blades are assesed
by comparing spectra with those stored for common conditions. Measurement
and analysis data are stored centrally and blade condition displayed using a web
browser.
Main components monitored: Blades
Monitoring technology: Accelerometer
Analysis method(s): FFT frequency domain
Data or sampling frequency: ≈ 1kHz

34 **FS2500: FiberSensing, Portugal**
Description: BraggSCOPE measurement unit designed for industrial
environments to interrogate up to four Fiber Bragg Grating sensors.
Acceleration, tilt, displacement, strain, temperature and pressure measurable.

Main components monitored: Blades
Monitoring technology: Fibre optic
Analysis method(s): Unknown
Data or sampling frequency: Up to 2kHz

35   **RMS (Rotor Monitoring System): Moog Insensys Ltd., UK**
Modular blade sensing system consisting of 18 sensors, six per blade, installed in the cylindrical root section of each blade to provide edgewise and flapwise bending moment data. Can be designed-in during turbine manufacture or retrofitted. Monitors turbine rotor performance, mass and aerodynamic imbalance, blade bending moments, icing, damage and lightning strikes. Possible integration, as an external input, in commercial available CMSs.
Main components monitored: Blades
Monitoring technology: Fibre optic strain
Analysis method(s): Time domain strain analysis
Data or sampling frequency: 25 Hz/sensor

36   **SCAIME Condition Monitoring Solutions: SCAIME, France**
Fibre optic systems for structural monitoring. Sensors, made of glass fibre reinforced plastic or aluminium alloys, measure the stresses on the blades, the mast and the foundations. MDX400 data acquisition unit with an integrated web server for remote system and sensor setup. Emergency alarms generated when loads become too high and blade loads data used for pitch controller input. Data processing provides remaining life estimation, defect and ice detection. In the mast, sensors measure bending moments at different heights to monitor tower deformations and oscillations. Sensors monitor foundation aging due to load accumulation, soil pressure, grouting.
Main components monitored: Blades, mast, foundation
Monitoring technology: Strain and temperature sensors; long base extensometers; displacement sensors; tilt-meters; accelerometers
Analysis method(s): Time domain analysis
Data or sampling frequency: 100 Hz

The first observation to make from Table 3.B1 is that the CMSs nearly all focus on the same WT subassemblies as follows:

- blades
- main bearing
- gearbox internals
- gearbox bearings
- generator bearings.

A quick summary of Table 3.B1 shows that there are:

- 27 systems primarily based on drive train vibration analysis (1–27)
- one system using motor current signature analysis, operational modal analysis and acoustic emission techniques (28)
- four systems solely for oil debris monitoring (29–32)
- one system using vibration analysis for WT blade monitoring (33)
- three systems based on fibre optic strain measurement in WT blades, mast and foundation (34, 35, 36)

The majority of the systems are based on monitoring methods originating from other, traditional rotating machinery industries. Indeed 27 of the 36 systems in the table are based on vibration monitoring using accelerometers typically using a configuration similar to that shown in Figure 3.B4 for the Mita-Teknik WP4086 CMS (27).

Of these 27 CMSs, all have the capability to carry out some form of diagnostic procedure once a fault has been detected. In most cases this is done through fast Fourier transform (FFT) analysis of high frequency data in order to detect fault-specific frequencies. In the case of the SKF WindCon 3.0 (23), the ACOEM OneProd Wind CMS (15) and several others, high data acquisition is triggered by operational parameters. For example, the SKF WindCon 3.0 CMS can be configured to collect a vibration spectrum on either a time basis or when a specific load and speed condition is achieved. The aim of this is to acquire data that is directly comparable between each point and, importantly, to allow spectra to be recorded in apparently stationary conditions. This is an important point to note when using traditional signal processing methods such as the FFT that require stationary signals in order to obtain a clear result. The Mita-Teknik WP4086 system (27), however, states that it includes advanced signal-processing techniques such as comb filtering, whitening and Kurtogram analysis which in combination with resampling and order alignment approaches, allow the system to overcome the effects of WT speed variations.

An innovative vibration-based CMS is OrtoSense APPA (2) which is based on auditory perceptual pulse analysis. This patented technology outperforms the human ear by capturing a detailed interference pattern and detecting even the smallest indication of damaged or worn elements within the machine/turbine. OrtoSense states that its product is four to ten times more sensitive compared to prevailing systems.

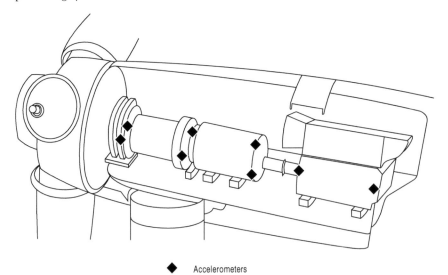

◆   Accelerometers

Figure 3.B4   Typical CMS accelerometer positions in the nacelle of a WT, adapted from (Isko et al., 2010)

CMSWind (7) is still in the development phase but it represents an advanced system for WT CM utilising three new and novel techniques, specifically designed for WTs and their components. Motor current signature analysis, operational modal analysis and acoustic emission techniques will be used to monitor the condition of the generator, the gearbox and rotary components, respectively. All systems will be tied together through SCADA to provide supervisory control, data logging and analysis.

Six of the vibration-based CMSs also state that they are able to monitor the level of debris particles in the WT gearbox lubrication oil system. Further to this, included in the table are four systems which are not in themselves CMSs. These four (29–32) are oil quality monitoring systems or transducers rather than full CMSs but are included, as discussion with industry has suggested that debris in oil plays a significant role in the damage and failure of gearbox components. Systems using these debris in oil transducers are using either cumulative particle counts or particle count rates.

Several of the 27 vibration-based CMSs also allow for other parameters to be recorded alongside vibration such as load, wind speed, generator speed and temperatures although the capabilities of some systems are unclear given the information available. There is some interest being shown as regards the importance of operational parameters in WT CM. This arises from the fact that many analysis techniques, for example the FFT, have been developed in constant speed, constant load environments. This can lead to difficulties when moving to the variable speed, variable load WT however experienced CM engineers are able to use these techniques and successfully detect faults.

Recent CM solutions, as (1), (7), (11), (13), (19), (24), (27), (28), (30), can be adapted and fully integrated with existing SCADA systems using standard protocols. Thanks to this integration, the analysis of the systems installed on the wind energy plant can also directly consider any other signals or variables of the entire controller network, as for example current performance and operating condition, without requiring a doubling of the sensor system. The database, integrated into a single unified plant operations' view, allows a trend analysis of the condition of the machine.

In some cases the CMS company offers also custom service solutions from 24/7 remote monitoring to on-demand technical support, examples are GE Energy ADAPT.wind (1), Moventas CMaS (10), ABS Wind Turbine In-Service (24) and several others.

Recently patented condition-based turbine health monitoring systems, as (4), (5), (13), (19), feature diagnostic and prognostic software unifying fleet wide CMS and SCADA enabling the identification of both source and cause of the fault and the application of prognostics to establish the remaining operational life of the component.

Three CMSs in the table (34, 35, 36) are based on strain measurement using fibre-optic transducers. FS2500 (34) and RMS (35) are aimed at detection of damage to WT blades and, in the case of the Moog Insensys system (35), blade icing, mass unbalance or lightning strikes. SCAIME system (36) allows turbine structural monitoring with sensors mounted on the blades, the mast and the foundations. These three systems may be fitted to WT retrospectively. Compared to vibration monitoring techniques, these systems can be operated at low sampling rates as they are looking to observe changes in time domain. They are usually integrated in the WT control system but there are also some cases of integration, as an external input, into commercial available conventional vibration-based CMSs. In addition to (34) (35) and (36) there is the IGUS system (33) that uses accelerometers to monitor blade damage, icing and lightning strikes. This system compares the blade accelerometer FFT with stored spectra for similar operating conditions and has the power to automatically shut down or restart a WT based on the results. The system appears to be popular within the industry.

Comments on numbers of CMS installed and centrally monitored

Brüel & Kjær Vibro

It has been reported that Brüel & Kjær had sold 4000 Vibro systems world-wide, all for WTs with 2500 connected to their central monitoring service. It has been reported that Bachmann, a WT controller manufacturer, is now a serious competitor. Brüel & Kjær take signals from CMS transducers and SCADA, as allowed by the WT OEM, which is straightforward with Vestas, where the Brüel & Kjær system is fitted to new turbines.

Gram & Juhl

It was reported that Gram & Juhl had 6000 CMS systems installed worldwide but not all in WTs and that 2000–3000 WT systems are connected to the Gram & Juhl monitoring centre. Gram & Juhl take signals from CMS transducers and SCADA, as permitted by the WT OEM, which is straightforward with Siemens, where the Gram & Juhl system is fitted to new turbines.

The process of automating CMS detection has proved very hard because it depends upon specific drive-train designs and required some learning of machine operation, that generally came from experienced CMS technicians.

Pruftechnik

Pruftechnik has more than 2000 systems installed in the wind industry with 800 WT systems connected to their monitoring centre. They are not selling directly to WT OEMs, except in rare cases, but did supply their system to gearbox OEMs, for example Winergy. They are also supplying an oil debris counter and handheld and alignment devices.

SKF Windcon

SKF have thousands of Windcon units fitted to WTs world-wide with about 1000 Windcons connected to their Hamburg Wind Centre.

It was reported that CMS is a difficult sell for WTs, as some operators refuse to recognise the value of CMS because it cannot prevent failure without interpretation. Their philosophy is run to failure. However, on large WFs in the US there is a growing interest as operators begin to recognise the disadvantage of simple availability benchmarks for operational performance measurement, where these can only be achieved at high O&M cost. Some operators, particularly of large WFs, are recognising the benefit of maintenance planning using integrated SCADA and CMS data.

Mita-Teknik

Mita-Teknik continue to offer their CMS option within their SCADA offering. The value of integration between SCADA and CMS was stressed and only Mita-Teknik appeared to offer that advantage to operators.

The future of wind turbine condition monitoring

As can be seen from this survey of current CMSs there is a clear trend towards vibration monitoring of WTs. This is presumably a result of the wealth of knowledge gained from many years work in other fields. It is likely that this trend will continue; however, it would be reasonable to assume that other CM and diagnostic techniques will be incorporated into existing systems.

Currently these additions are those such as oil debris monitoring and fibre optic strain measurement. However, it is likely that major innovation will occur in terms of developing signal processing techniques. In particular, the industry is already noting the importance of operational parameters such as load and speed and so techniques may begin to adapt further to the WT environment leading to more reliable CMSs, diagnostics and alarm signals.

Automation of CM and diagnostic systems may also be an important development as WT operators acquire a larger number of turbines and manual inspection of data becomes impractical. Further to this, it is therefore essential that methods for reliable and automatic diagnosis are developed with consideration of multiple signals in order to improve detection and increase operator confidence in alarm signals. It is clear that CMS automation is difficult, because of individual plant peculiarities, but that with larger WFs it is becoming more attractive for operators.

However, it should be noted that a major hindrance to the development of CMSs and diagnostic techniques could be data confidentially, which means that few operators are able to divulge or obtain information concerning their own WTs. This is an issue which should be addressed if the art of CM is to progress quickly. Confidentiality has also led to a lack of publicly available cost justification of WT CM, which seems likely to provide overwhelming support for WT CM, particularly in the offshore environment where availability is at a premium.

Conclusions

From this survey we can conclude that:

- current WT reliability is reasonable however in the offshore environment the failure rate will be unacceptable;
- cost effective and reliable CM is required to enable planned maintenance, reduce unplanned WT downtime and improve capacity factors;
- successful CMSs must be able to adapt to the non-stationary, variable speed nature of WTs;
- there is a wide variety of commercially available CMSs currently in use on operational WTs;
- monitoring technology is currently based on techniques from other, conventional rotating machine industries;
- vibration monitoring is currently favoured in commercially available systems using standard time and frequency domain techniques for analysis;
- these traditional techniques can be applied to detect WT faults but require experienced CM engineers for successful data analysis and diagnosis;
- some commercially available CMSs are beginning to adapt to the WT environment and to be fully integrated into existing SCADA systems;
- recently patented condition-based turbine health monitoring systems feature diagnostic and prognostic software enabling the identification of both source and cause of the fault and the application of prognostics to establish the remaining operational life of the component;
- a diverse range of new or developing technologies are moving into the WT CM market.

Finally, it should be noted that there is not currently a consensus in the WT industry as to the correct route forward for CM of WTs. Work in this document and its references suggest that CM of WTs will be beneficial for large onshore WTs but essential for all offshore development and should be considered carefully by the industry as a whole.

References

Amirat, Y., Benbouzid, M.E.H., Bensaker, B. & Wamkeue, R. 2007. Condition Monitoring and Fault Diagnosis in Wind Energy Conversion Systems: A Review. Proceedings of IEEE International Electric Machines & Drives Conference. IEMDC '07 May 3–5 2007. pp. 1434–1439.

Amirat, Y., Benbouzid, M.E.H., Al-Ahmar, E., Bensaker, B. & Turri, S. 2009. A Brief Status on Condition Monitoring and Fault Diagnosis in Wind Energy Conversion Systems. *Renewable and Sustainable Energy Reviews*, 13(9), pp. 2629–2636.

Besnard, F. 2013. On Maintenance Optimization for Offshore Wind Farms. KTH Royal Institute of Technology School of Electrical Engineering, Division of Electromagnetic Engineering, Stockholm, Sweden. PhD Thesis.

Besnard, F., Nilsson, J. & Bertling, L. 2010. On the Economic Benefits of Using Condition Monitoring Systems for Maintenance Management of Wind Power Systems.

Proceedings of 2010 IEEE 11th International Conference on Probabilistic Methods Applied to Power Systems (PMAPS), Singapore.

Chen, B., Zappalá, D., Crabtree, C.J. & Tavner, P.J. 2014. Survey of Commercially Available SCADA Data Analysis Tools for Wind Turbine Health Monitoring. Durham University and the SUPERGEN Wind Energy Technologies Consortium, available at http://dro.dur.ac.uk/12563/ and http://www.supergen-wind.org.uk/dissemination.html

Chen, B, Matthews, P.C., Tavner, P.J. 2015. Automated On-Line Fault Prognosis for Wind Turbine Pitch Systems using Supervisory Control and Data Acquisition, IET, Renewable Power Generation, doi: 10.1049/iet-rpg.2014.0181

Ciang, C.C., Lee, J.-R. & Bang, H.-J. 2008. Structural Health Monitoring for a Wind Turbine System: a Review of Damage Detection Methods. *Measurement Science and Technology*, 19(12), pp. 1–20.

Crabtree, C.J. 2011. Condition Monitoring Techniques for Wind Turbines. Durham University, UK, PhD Thesis.

Crabtree, C.J. 2012. Operational and Reliability Analysis of Offshore Wind Farms. Proceedings of the Scientific Track of the European Wind Energy Association Conference, April 16–19, 2012, Copenhagen, Denmark.

Crabtree, C.J., Zappalá, D. & Tavner, P.J. 2014. Survey of Commercially Available Condition Monitoring Systems for Wind Turbines. Durham University and the SUPERGEN Wind Energy Technologies Consortium. available at http://dro.dur.ac.uk/12497/ and http://www.supergen-wind.org.uk/dissemination.html.

Djurović, S., Crabtree, C.J., Tavner, P.J. & Smith, A.C. 2012. Condition Monitoring of Wind Turbine Induction Generators with Rotor Electrical Asymmetry. *IET Renewable Power Generation*, 6(4), pp. 207–216.

Doner, S. 2009. *Winergy Condition Diagnostics System Enterprise-Wide Fleet Management Roadmap to Advanced Condition Monitoring.* Broomfield, CO: NREL Wind Turbine Condition Monitoring Workshop, available at http://wind.nrel.gov/public/Wind_Turbine_Condition_Monitoring_Workshop_2009/

Eggersglüß, W. 1995–2004. Wind Energie IX–XIV, Praxis-Ergebnisse 1995–2004. Landwirtschaftskammer Schleswig-Holstein, Osterrönfeld, Germany.

Faulstich, S., Durstewitz, M., Hahn, B., Knorr, K. & Rohrig, K. 2008. *German Wind Energy Report 2008.* Kassel, Germany: Institut für Solare Energieversorgungstechnik, available at http://windmonitor.iwes.fraunhofer.de/wind/download/Windenergie_Report_2008_en.pdf

Faulstich, S., Hahn, B. & Tavner, P.J. 2011. Wind Turbine Downtime and Its Importance for Offshore Deployment. *Wind Energy*, 14, pp. 327–337.

Feng, Y., Tavner, P.J. & Long, H. 2010. Early Experiences with UK Round 1 Offshore Wind Farms. *Proceedings of the Institution of Civil Engineers: energy*, 163(4), pp. 167–181.

Feng, Y., Qiu, Y., Crabtree, C.J., Long, H. & Tavner, P.J. 2013. Monitoring Wind Turbine Gearboxes. *Wind Energy*, 16, pp. 728–740.

Fischer, K. 2012. Maintenance Management of Wind Power Systems by means of Reliability-Centred Maintenance and Condition Monitoring Systems. Chalmers University, available at http://www.chalmers.se/en/projects/Pages/Maintenance-Management-of-Wind-Power-Systems.aspx

Garcia Marquez, F.P., Tobias, A.M., Pinar Perez, J.M. & Papaelias, M. 2012. Condition Monitoring of Wind Turbines: Techniques and Methods. *Renewable Energy*, 46, pp. 169–178.

Germanischer Lloyd 2003. Guideline for the Certification of Wind Turbines, Edition 2003 with Supplement 2004.. Hamburg, Germany.

Germanischer Lloyd 2005. Guideline for the Certification of Offshore Wind Turbines, Edition 2005. Hamburg, Germany.

Germanischer Lloyd 2007. Guideline for the Certification of Condition Monitoring Systems for Wind Turbines. Hamburg, Germany.

Giordano, F. & Stein, R. 2011. The Operator's Assessment of Condition Monitoring: Practical Experience and Results. Proceedings of the European Wind Energy Association Conference, Brussels, Belgium.

Gram & Juhl A/S 2010. TCM® Turbine Condition Monitoring, available at http://www.rotomech.com/pdf/brochure-eng.pdf

Gray, C.S. & Watson, S.J. 2010. Physics of Failure Approach to Wind Turbine Condition Based Maintenance. *Wind Energy*, 13(5), pp. 395–405.

Hahn, B., Durstewitz, M. & Rohrig, K. 2007. Reliability of Wind Turbines, Experience of 15 years with 1,500 WTs. Wind Energy – Proceedings of the Euromech Colloquium, Springer, Berlin, Germany.

Hameed, Z., Hong, Y.S., Cho, Y.M., Ahn, S.H. & Song, C.K. 2009. Condition Monitoring and Fault Detection of Wind Turbines and Related Algorithms: A Review. *Renewable and Sustainable Energy Reviews*, 13(1), pp. 1–39.

Hyers, R.W., Mcgowan, J.G., Sullivan, K.L., Manwell, J.F. & Syrett, B.C. 2006. Condition Monitoring and Prognosis of Utility Scale Wind Turbines. *Energy Materials: Materials Science and Engineering for Energy Systems*, 1(3), pp. 187–203.

Isko, V., Mykhaylyshyn, V., Moroz, I., Ivanchenko, O. & Rasmussen, P. 2010 Remote Wind Turbine Generator Condition Monitoring with WP4086 System, Materials Proceedings, European Wind Energy Conference 2010, Warsaw, Poland.

ISO 14224:2006 Petroleum, Petrochemical and Natural Gas Industries – Collection and Exchange of Reliability and Maintenance Data for Equipment.

Lu, B., Li, Y., Wu, X. & Yang, Z. 2009. A Review of Recent Advances in Wind Turbine Condition Monitoring and Fault Diagnosis. Proceedings of 2009 IEEE Symposium on Power Electronics and Machines in Wind Applications. PEMWA, June 24–26, 2009, Lincoln, NE, pp. 1–7.

Mcgowin, C., Walford, C. & Roberts, D. 2006. Condition Monitoring of Wind Turbines – Technology Overview, Seeded-Fault Testing and Cost-Benefit Analysis. EPRI, available at http://www.epri.com/abstracts/Pages/ProductAbstract.aspx?ProductId=000000000001010419

Nilsson, J. & Bertling, L. 2007. Maintenance Management of Wind Power Systems Using Condition Monitoring Systems – Life Cycle Cost Analysis for Two Case Studies. *IEEE Transactions on Energy Conversion*, 22(1), pp. 223–229.

Noordzeewind, S. 2007–2009. Egmond aan Zee Operations Reports 2007, 2008, 2009. www.noordzeewind.nl.

Pinar Pérez, J.M., García Márquez, F.P., Tobias, A. & Papaelias, M. 2013. Wind Turbine Reliability Analysis. *Renewable and Sustainable Energy Reviews* 23, pp. 463–472.

Qiu, Y., Feng, Y., Tavner, P., Richardson, P., Erdos, G. & Chen, B. 2012. Wind Turbine SCADA Alarm Analysis for Improving Reliability. *Wind Energy*, 15(8), pp. 951–966.

Randall, R.B. 2011. *Vibration-Based Condition Monitoring: Industrial, Aerospace and Automotive Applications*, John Wiley & Sons.

Reliawind 2011. Project Final Report, available at http://cordis.europa.eu/publication/rcn/14854_en.html.

Ribrant, J. & Bertling, L.M. 2007. Survey of Failures in Wind Power Systems with Focus on Swedish Wind Power Plants During 1997–2005. *IEEE Transactions on Energy Conversion*, 22(1), pp. 167–173.

Sheng, S. & Veers, P. 2011. Wind Turbine Drivetrain Condition Monitoring – An Overview. Proceedings of Machinery Failure Prevention Technology (MFPT) Society 2011 Conference, May 10–12, 2011, Virginia Beach, VA.

Spinato, F., Tavner, P.J., Van Bussel, G.J.W. & Koutoulakos, E. 2009. Reliability of Wind Turbine Subassemblies. *IET Renewable Power Generation*, 3(4), pp. 387–401.

Tavner, P.J. 2012. *Offshore Wind Turbines: Reliability, Availability and Maintenance*, London, The Institution of Engineering and Technology.

Tavner, P.J., Xiang, J. & Spinato, F. 2007. Reliability Analysis for Wind Turbines. *Wind Energy*, 10(1), pp. 1–18.

Tavner, P.J., Ran, L., Penman, J. & Sedding, H. 2008. *Condition Monitoring of Rotating Electrical Machines*, London, The Institution of Engineering and Technology.

Tavner, P., Faulstich, S., Hahn, B. & Van Bussel, G.J.W. 2010. Reliability & Availability of Wind Turbine Electrical & Electronic Components. *EPE Journal*, 20(4), pp. 45–50.

Walford, C.A. 2006. Wind Turbine Reliability: Understanding and Minimizing Wind Turbine Operation and Maintenance Costs. Sandia National Laboratories., SAND2006–1100, available at http://prod.sandia.gov/techlib/access-control.cgi/2006/061100.pdf.

Watson, S.J., Xiang, B.J., Yang, W.X., Tavner, P.J. & Crabtree, C.J. 2010. Condition Monitoring of the Power Output of Wind Turbine Generators Using Wavelets. *IEEE Transactions on Energy Conversion*, 25(3), pp. 715–721.

Wiggelinkhuizen, E., Verbruggen, T., Braam, H., Rademakers, L., Xiang, J.P. & Watson, S. 2008. Assessment of Condition Monitoring Techniques for Offshore Wind Farms. *Transactions of the ASME Journal of Solar Energy Engineering*, 130(3), p. 9.

Wilkinson, M. 2011. Measuring Wind Turbine Reliability – Results of the Reliawind Project. Proceedings of the Scientific Track of the European Wind Energy Association Conference, Brussels, Belgium.

Wilkinson, M., Spinato, F., Knowles, M. & Tavner, P.J. 2006. Towards the Zero Maintenance Wind Turbine. Proceedings of the 41st International Universities Power Engineering Conference, Vol 1. pp. 74–78, Newcastle, UK.

Wilkinson, M.R., Spinato, F. & Tavner, P.J. 2007. Condition Monitoring of Generators & Other Subassemblies in Wind Turbine Drive Trains. Proceedings of 2007 IEEE International Symposium on Diagnostics for Electric Machines, Power Electronics and Drives (SDEMPED 2007), Cracow, Poland. pp. 78–82.

Wilkinson, M., Hendriks, B., Spinato, F., Gomez, E., Bulacio, H., Roca, J., Tavner, P.J., Feng, Y. & Long, H. 2010. Methodology and Results of the ReliaWind Reliability Field Study. Proceedings of the Scientific Track of the European Wind Energy Association Conference, Warsaw, Poland.

Wilkinson, M., Harman, K., Hendriks, B., Spinato, F. & Van Delft, T. 2011. Measuring Wind Turbine Reliability – Results of the Reliawind Project. Proceedings of the Scientific Track of the European Wind Energy Association Conference, Brussels, Belgium.

Wind on the Grid 2009, available at http://www.windgrid.eu/Deliverables_EC/D6%20 WCMS.pdf

Yang, W., Tavner, P.J. & Crabtree, C.J. 2009a. An Intelligent Approach to the Condition Monitoring of Large Scale Wind Turbines. Proceedings of the Scientific Track of the European Wind Energy Association Conference, 2009 Marseille, France.

Yang, W., Tavner, P.J. & Wilkinson, M.R. 2009b. Condition Monitoring and Fault Diagnosis of a Wind Turbine Synchronous Generator Drive Train. *IET Renewable Power Generation*, 3(1), pp. 1–11.

Yang, W., Tavner, P.J., Crabtree, C.J. & Wilkinson, M. 2010. Cost-Effective Condition Monitoring for Wind Turbines. *IEEE Transactions on Industrial Electronics*, 57(1), pp. 263–271.

Yang, W., Tavner, P.J., Crabtree, C.J., Feng, Y. & Qiu, Y. 2014. Wind Turbine Condition Monitoring: Technical and Commercial Challenges. *Wind Energy*, 17(5), pp. 673–693.

Zaher, A., Mcarthur, S.D.J., Infield, D.G. & Patel, Y. 2009. Online Wind Turbine Fault Detection Through Automated SCADA Data Analysis. *Wind Energy*, 12(6), pp. 574–593.

Zappalá, D., Tavner, P.J., Crabtree, C.J. & Sheng, S. 2014. Side-Band Algorithm for Automatic Wind Turbine Gearbox Fault Detection and Diagnosis. *IET Renewable Power Generation*, 8(4), pp. 380–389.

4 Turbine blade materials and modelling

Geoff Dutton

Introduction

Composite materials and structures have been developed extensively over the last three decades and used widely in, for example, bridges, aerospace, and wind turbines. Modern wind turbine blades are designed almost exclusively with fibre-reinforced composite materials due to their superior strength and stiffness per unit mass. In addition to high specific strength and specific stiffness, high quality composite materials possess superior fatigue resistance compared with most conventional metals.

The usual effect of fatigue in metals at low stresses is simply to harden the metal slightly. Fatigue in metals often progresses by the initiation of a single crack and its intermittent propagation until catastrophic failure which occurs with little warning. Unlike metals, fibre-reinforced composite materials are inhomogeneous and anisotropic; they accumulate damage in a general rather than a localised fashion, and failure does not generally occur by the propagation of a single macroscopic crack (Harris, 2003). Fatigue damage in a composite material is distributed throughout the stressed region, often resulting in reduced stiffness before any major reduction in strength. The micro-structural mechanisms of damage accumulation, including fibre breakage and matrix cracking, de-bonding, transverse-ply cracking, and delamination, occur sometimes independently and sometimes interactively and their relative importance may be strongly affected by both material properties and testing conditions (Harris, 2003).

A rotating wind turbine blade is subject to many forces (including gyroscopic, gravity, centrifugal, steady and unsteady aerodynamic) which make the fatigue design for the rotor and overall structure very complex. In fact, a wind turbine blade typically endures several orders of magnitude more cycles than an aircraft. The design philosophy for wind turbine blades emphasises adequate strength and stiffness for minimum weight and cost, resulting in the use of significantly different materials than in the safety-critical aerospace industry.

Extensive fundamental materials testing has been carried out under the auspices of the EC projects Optimat Blades (Janssen et al., 2006) and UPWIND (EWEA, 2011). The focus within SUPERGEN Wind has been on multi-mode testing and the join between materials, particularly with respect to the attachment of the shear web to the blade skin. The T-joint was taken as representative of the join between shear web and the blade skin. A particular interest was in the potential to use stitched bonded joints, or even fully 3D fabrics, to improve the pull-out toughness.

The following section deals with the basic materials used in the study. The performance of these materials in a T-joint test coupon is then discussed. The final section develops a fully parametric blade finite element model able to subject these innovative material concepts to a typical wind turbine blade loading.

Basic materials

The most common wind turbine blade structure comprises a load-carrying blade skin, usually laid up in two halves, with one or two internal shear webs to carry the shear loads between the pressure-side and suction-side skins. A possible alternative arrangement is to replace the shear webs with a load-carrying spar bonded to two thinner, lighter blade skins. Whichever construction is favoured, key points of weakness lie in sharp changes of geometry (where fabric is stretched or constricted resulting in resin-rich or resin-poor areas) and the join between structural elements, such as the shear web or spar and the blade skin. While fibre composite materials have very good in-plane mechanical properties, they are typically much weaker in the through-thickness direction because the load in this case is predominantly carried by the resin matrix.

The SUPERGEN Wind project has looked at the possibility of reinforcing joins or weaker blade areas in the through-thickness direction using techniques ranging from interlaminar veils, through out-of-plane stitching, to the use of fully 3D woven fabrics. The individual material components are summarised below and results presented for their basic strength and fatigue properties when laid up in a conventional rectangular test coupon and tested in uniaxial quasi-static tension and tension-tension fatigue. The incorporation of these materials into the design and manufacture of the T-joint test coupon will be introduced at the start of the following section, followed by a full discussion of static and fatigue test results, and the development of models to predict the observed failure behaviour.

Matrix

The most widely used matrix materials in the wind turbine blade industry are thermoset polymer resins such as polyester and epoxy. Desirable properties are low viscosity and relatively long gel time to allow time for infusion, alongside high ductility and toughness to restrict crack propagation. In this work an epoxy resin based on Araldite LY 564 and a formulated amine hardener based on XB 3486, both supplied by Huntsman, have been used. Manufacturer's data for resin and hardener are listed in Table 4.1.

Table 4.1 Resin and hardener

	Viscosity @ 25oC (mPas)	Density @ 25oC (g/cm3)	Flash point (°C)
Araldite LY 564	1200–1400	1.1–1.2	185
XB 3484	10–20	0.94–0.95	123
Albipox 3001	22000	1.11–1.14	150
Albipox F091	17000	1.15–1.17	150
Albidur XP 1/609	7000	–	>150

Fabrics

Two different types of glass fibre fabric were used in this experimental campaign: unidirectional 0° glass fibre (FGE664) as stiffener and 45°glass fibres (FGE104) to manufacture the main part of the T-joint. Both of these fabrics were manufactured by Formax. Details of all fabrics used in this project are listed in Table 4.2.

Interleaf veils

A non-woven tissue fabric was used in composites in order to increase interlaminar fracture toughness. Details of these randomly oriented fabric layers are summarised in Table 4.3.

Tufting material

Out-of-plane stitching was carried out with 120 tex Kevlar-29 thread with tenacity of 185–200 cN/tex from Atlantic Thread and Supply.

Table 4.2 Fabric specification

	Structure	Fibre type	Fabric weight (g/m²)
FGE104	±45	E-Glass	609 ± 5%
FGE664	UD	E-Glass	594 ± 5%
38399	Plain weave	Carbon-6K	375 ± 5%

Table 4.3 Interleaf veil data

Code	Material	Thickness (mm)	Areal weigh (g/m2)	Fibre diameter (µm)
PA	Polyamide	0.194	28	12
PE	Polyester	0.055	12	15

3D fibre weaves

3D fabrics were selected to manufacture the composite T-joints in order to study the effect of out-of-plane reinforcement in the web. Two types of 3D architecture were adopted: angle interlock structure and layer-to-layer stitched structure. The structure was woven as a flat panel, then the middle layer in the through-thickness direction was removed to produce a gap in thickness. The woven panel was then opened up from this gap and the two flange sections of the joint formed on each side of the web. The fibre type used in the 3D weaves was E-glass.

Characterisation of stitched test coupons under quasi-static and fatigue loading

A major potential problem of stitching is that localised damage can occur where the sewing needle and yarn penetrate the main structural materials. In order to characterise the damage phenomena more clearly, a set of uniaxial quasi-static tension and tension-tension fatigue tests were carried out on both baseline and stitched test coupons, monitored using a range of non-destructive testing (NDT) techniques: digital image correlation (DIC) to measure surface contours and three-dimensional displacement/strain fields; infra-red thermography (IRT) to measure temperature increase under fatigue loading; and x-ray tomography (XRT) to show damage development.

The test coupons were manufactured using eight layers of Uniweave glass fabric supplied by Carr Reinforcements Ltd. This fabric has a construction in which the unidirectional (UD) fibres are bonded together in the weft direction by a very light polyester thread resulting in negligible crimp. Modified lock stitches of 120tex Kevlar-29 thread (see above) were applied into the fabric layup using an industrial sewing machine in five parallel stitching lines. Vacuum-assisted resin transfer moulding (VARTM) was used to manufacture each composite plate, which was then weighed and cut into coupons with the required dimensions using a water-cooled diamond saw. A typical stitched coupon sample is shown in Figure 4.1. The dimensions and volume fraction of fibres for both stitched and unstitched samples are listed in Table 4.4.

ASTM standards D3039-08 and D3479-96 were followed for measuring quasi-static tensile and tension-tension fatigue properties, respectively. A constant stress ratio ($R = 0.1$) and frequency (5 Hz) were applied to all fatigue testing.

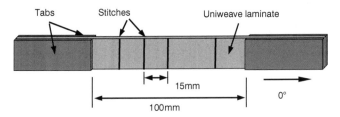

Figure 4.1 Stitched laminate test coupon geometry

Table 4.4 Stitched/unstitched coupon dimensions

Sample	Width (mm)	Thickness (mm)	Fibre volume fraction (%)
Unstitched	15.42 ± 0.49	4.98 ± 0.13	55.8
Stitched	20.07 ± 0.11	5.22 ± 0.12	55.9

In the unstitched test coupons, damage initiated randomly throughout the sample, but for the stitched coupons damage initiated at the stitches. This difference was confirmed by both the DIC and IRT measurements (see below). However, despite different initiation patterns, both unstitched and stitched samples failed in the same way by splitting and fracturing of fibre tows. The S-n curves for the two types of coupon were very similar (Figure 4.2).

Digital image correlation

A DIC system Q-400 developed by Dantec Dynamics was used to track the strain development at the sample surface. Measurements were taken by pausing the fatigue load at the minimum tensile stress after a given number of cycles, and then quasi-

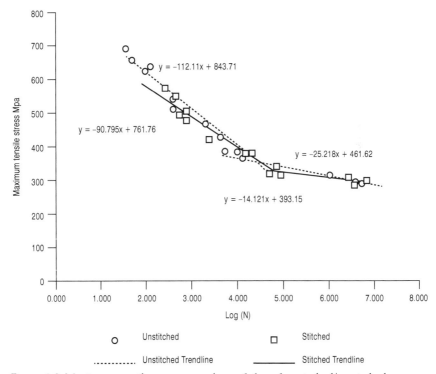

Figure 4.2 Maximum tensile stress v. cycles-to-failure for stitched/unstitched coupons under tension-tension fatigue

statically increasing the load up to the maximum tensile stress. These tests showed that the through-thickness stitching sites generate strain concentrations within the coupon. Figure 4.3 shows results along two interrogation lines for the stitched coupon, one on the stitching line and the other equidistant between stitching lines. The mean strain variation is also included in the figure for both the stitched and unstitched coupons. The mean strain levels in the two coupon types are similar, but the strain along the stitching line increases up to 50 per cent higher than in unstitched areas; it was also seen to increase gradually during the fatigue test.

Infrared thermography

It is well known that when a material is subjected to a change in mechanical loading, its temperature also changes. Typically, there will be both a reversible adiabatic temperature change due to the thermoelastic effect (Dulieu-Barton, 1999) and, for cyclic loading, a long-term increase in temperature due to cumulative viscoelastic energy dissipation within the material. These effects are elevated by stress concentration so that both instantaneous stress maxima and accumulating damage can be detected.

For the current study, a Thermosensorik infra-red camera operating in the 3–5μm wavelength band was used in thermographic mode to track temperature changes throughout the fatigue tests on stitched and unstitched coupons. Figure 4.4 shows the local temperature increase captured at various stages of a single load cycle in the middle fatigue life for a stitched coupon (the infra-red

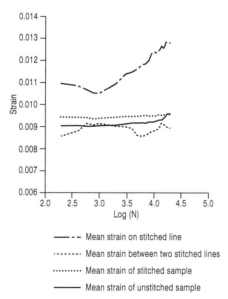

Figure 4.3 Principal strain DIC measurement – variation of mean strain along interrogation lines 1 and 2 for stitched coupon compared with mean strain for unstitched coupon

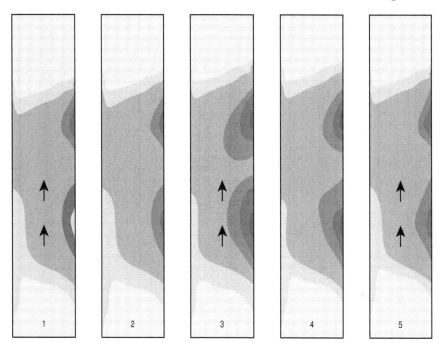

Figure 4.4 IRT images taken at five instances during a single mid-term fatigue cycle for the stitched sample (1 and 5 are minimum tensile stress, 3 is maximum tensile stress)

image collection frequency allowed 7.75 images per fatigue cycle at the loading frequency of 5 Hz). The highest temperature (white) occurs at the ends of the test piece, indicating frictional heating between the sample and the grips; the temperature distribution in the centre of the sample (similar for both stitched/ unstitched) is asymmetrical across the width of the test piece, probably arising from a matching asymmetry in the heat coming from the grips; two lines of stitches are visible (arrowed) indicating higher stress concentration and possibly also increased energy dissipation.

X-ray tomography

Thanks to the art of reconstructing a sliceable virtual 3D copy of an object from two-dimensional (2D) images, x-ray tomography is becoming increasingly popular (Withers, 2007). In contrast with the other two NDT techniques discussed here, detailed characterisation of fatigue crack mechanisms inside a material becomes possible with x-ray tomography. To date, no comprehensive study has been published for composite materials.

X-ray tomography is a non-destructive technique based on the absorption of x-rays as they pass through an object. Laboratory x-ray tomography systems typically utilise cone-beam geometry, in which x-rays are produced from a small spot in the source, irradiating the sample mounted on a rotation stage, so that the transmitted x-rays can be recorded against a pixellated detector. Geometric magnification is used to achieve high resolution, which is ultimately limited by the spot-size of the source. During a scan, a series of radiographs are taken as the object rotates over 360 degrees. A 3D reconstruction of the object is then obtained by using the Feldkamp–Davis–Kress algorithm.

For the current study, scans were performed using a Nikon Metrology XTH 225 system based on a 225 kV source, having a ~5 micron spot size and a choice of target material, including W, Cu, Mo and Ag. The system has a 14-bit flat panel detector with a 130 micron pixel pitch.

Observations with both DIC and IRT have indicated that the locations of stitches are potential areas for initiating failure in the stitched coupons, but the XRT analysis suggests that interlaminar cracks and delaminations (as must also be the case for the unstitched coupons) actually tend to develop in other locations separate from the stitches.

T-joints

The aim of material development for wind turbine blades is to find materials capable of extending the blade's working life and minimising geometry-related fatigue issues. To achieve this goal a T-section test coupon was designed as a representative element of the blade's shear web to skin connection. T-sections were made of glass fabric infused with epoxy resin using a vacuum assisted resin transfer moulding technique. The structure was modified with different toughening techniques to increase its interlaminar fracture toughness (i.e. resistance to crack initiation and propagation) with the view to delaying or stopping failure processes. These techniques included the use of veil layers, tufting and 3D weaving techniques.

Geometry, layup, and manufacture

The geometry and layup of the composite T-joint is shown in Figure 4.5. The width of the specimen is 50 mm. The layup comprises 5 layers of unidirectional glass fabric to represent the outer blade skin with 5 layers of ±45 glass fabric for each shear web flange, resulting in 10 layers of fabric within the web.

All T-section samples were manufactured using Vacuum Assisted Resin Infusion, a method widely used in the wind turbine industry for blade manufacturing.

To identify the mechanical properties of the T-joints with different design and modifications, T-joint specimens were tested under static and dynamic loading conditions to obtain the ultimate failure strength and fatigue performance. The various specimen configurations and details such as fibre type, matrix material and layup architecture are listed in Table 4.5.

The test coupon with polyamide veil was laid up in the same way but with a polyamide veil layer added between the L-shaped fabric section and the flat outer skin fabric section (Figure 4.6(b)). Observation of the baseline coupon

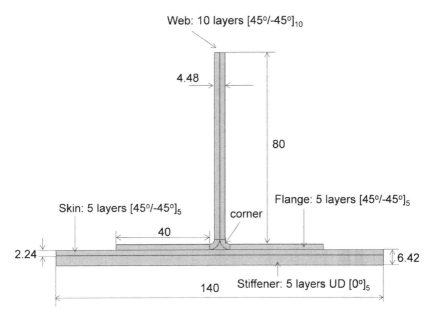

Figure 4.5 Geometry and layup of composite T-joint

Table 4.5 T-joint lay configurations and layups

Code	Fibre	Matrix	Weave	Architecture	Veil	Tufting
Base	Glass	Epoxy (LY564)	Not woven	Flange [±45]10[0]5 Web [±45]10	N/A	N/A
PA	Glass	Epoxy (LY564)	Not woven	Flange [±45]5PA[±45]5[0]5 Web [±45]10	Polyamide	N/A
PE	Glass	Epoxy (LY564)	Not woven	Flange [±45]5PE[±45]5[0]5 Web [±45]10	Polyester	N/A
T	Glass	Epoxy (LY564)	Not woven	Flange [±45]10[0]5 Web [±45]10	N/A	Flange-skin
2T	Glass	Epoxy (LY564)	Not woven	Flange [±45]10[0]5 Web [±45]10	N/A	Flange-skin-web
3D-4L	Glass	Epoxy (LY564)	3D woven layer to layer stitched	Flange 3D-4L[0]5 Web 3D-4L	N/A	N/A
3D-An	Glass	Epoxy (LY564)	3D woven angle interlocked	Flange 3D-An[0]5 Web 3D-An	N/A	N/A

fatigue tests showed that the crack initiates and propagates in this region and the hypothesis was that the polyamide veil would increase the toughness of the sample, thereby extending the fatigue life of the specimen.

The tufted test coupon was laid up with the same stack of fabric as for the baseline, reinforced with Kevlar thread in the through-thickness direction. Stitch density was 10 mm in the length direction and 10 mm in the width direction. The tufting was performed using a 2-axis robotic machine (Figure 4.7).

All coupons were manufactured by resin infusion (Figure 4.8), using a mould and airtight bag with the inlet connected to a resin container and the outlet to

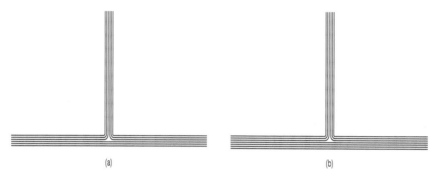

(a)   (b)

Figure 4.6 Schematic layup for T-section coupons: (a) baseline and (b) with veil

Figure 4.7 Tufting process

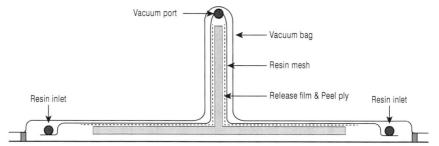

Figure 4.8 Resin infusion moulding of T-joint test coupons

a vacuum pump. To reduce the number of defects, the resin was degassed for 90 minutes at 25°C prior to infusion. After infusion under vacuum, the mould was transferred to an oven for curing. After curing the samples were removed from the mould and inspected structurally using a Midas NDT Jet Probe ultrasonic scan with 10 MHz probe.

Test procedure: Static test

Static tensile (pull-out) tests were performed on the T-joint test coupons to determine the maximum load carried by specimens before failure. The pull-out strength was measured using an Instron 5982 electro-mechanical testing machine with hydraulic grips set to a grip pressure of 50 bar. The machine was operated in displacement control at a displacement rate of 2 mm/min. Load and displacement values were measured using the test machine's load cell and jaw movement sensor respectively. Failure criteria to stop the test were determined as either 40 per cent drop in the measured load or development of a crack with total length greater than 50 mm in the region between flange and skin of the T-joint (i.e. 25 mm on each side of the web).

Test procedure: Fatigue test

Tension-tension ($R = 0.1$) fatigue tests were performed on the T-joint test coupons to develop comparable S-n curves. The tests were carried out using an Instron 8802 servo hydraulic fatigue machine equipped with hydraulic grips, again set to a grip pressure of 50 bar.

Load levels for the fatigue tests were chosen with regard to the maximum pull-out strength obtained from the static tensile (pull-out) tests, namely 40 per cent, 50 per cent, 60 per cent, 70 per cent and 80 per cent of the maximum pull-out strength. The test frequency was limited to 6 Hz to avoid an excessive temperature increase which might have resulted in changes to the mechanical properties of the composite.

The T-joint coupons were painted and gauged on one side to provide sufficient contrast for measuring crack length. Failure criteria for fatigue tests were defined as crack length of 25 mm in flange and skin bond line on each side of the web (total crack length of 50 mm) or reaching one million cycles in fatigue test. A video camera was used to measure the crack length and hence propagation rate throughout each test.

Ultimate strength of composite T-joints

The pull-out test results were normalised by the web cross-section area in order to remove the effect of coupon geometry. The ultimate strength of the section was therefore defined as:

$$\sigma_{strength} = \frac{\text{Applied load}}{\text{Area of web cross-section}}.$$

Load-displacement curves are presented here for the baseline coupons (Figure 4.9) and 3D woven layer-to-layer stitched (Figure 4.10). These show good repeatability of the test, reduced joint stiffness for the 3D woven joint configuration, and higher overall strength but with less resistance to failure after initiation.

Figure 4.11 shows the ultimate static strength for all the T-joint configurations tested. The 3D woven T-joints have the highest pull-out strength, followed by the polyester veil (which performs considerably better than the polyamide veil), and then the tufted coupons.

Fatigue performance of composite T-joints

Figure 4.12 shows the s-N curves for all T-joint configurations tested. The 3D woven T-joint with angle interlock weave structure exhibited almost 40 per cent improvement in fatigue life compared with the Baseline design. The best tufted coupon results showed a similar level of improvement, but the scatter amongst these results was much higher than for the 3D woven.

Finite element analysis of failure of composite T-joints

A finite element model of the baseline composite T-joints was constructed using the Abaqus program (Figure 4.13 and Figure 4.14) and used as an interpretation tool for the experimental results. Solid elements are used throughout, incorporating the layup shown in Figure 4.5.

Cohesive surface behaviour is used to model delamination of composite T-joints. Linear elastic traction-separation behaviour is used to model interface where the delamination occurs. The traction-separation model in Abaqus assumes initially linear elastic behaviour followed by the initiation and evolution of damage. Damage is assumed to initiate when the maximum nominal stress ratio reaches a value of one. Interlaminar strength is introduced as a variable parameter in order to investigate its effect on the initiation of delamination.

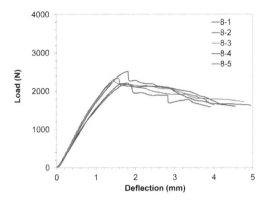

Figure 4.9 Baseline T-joints – load-displacement behaviour under quasi-static load

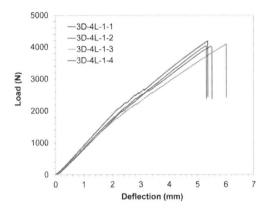

Figure 4.10 3D woven layer-to-layer stitched T-joints – load-displacement behaviour under quasi-static load

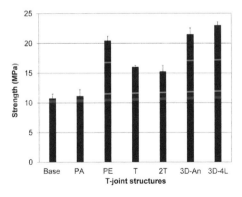

Figure 4.11 Ultimate pull-out strength of various T-joint configurations (for key to abbreviations see Table 4.5)

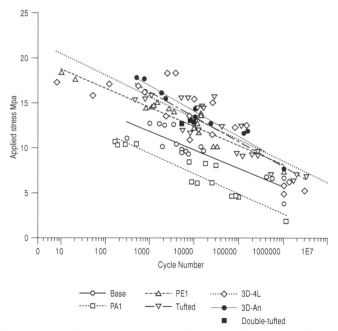

Figure 4.12 S-n curves for various T-joint configurations (for key to abbreviations see Table 4.5)

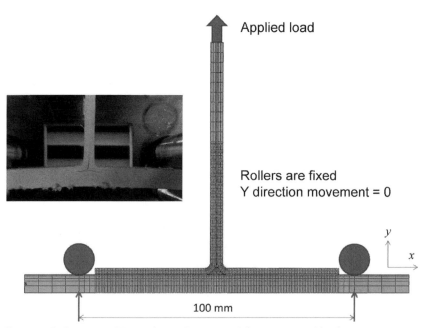

Figure 4.13 Composite T-joint finite element model geometry and loading

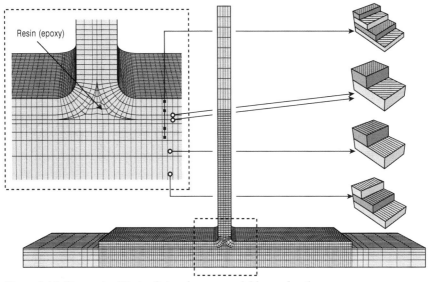

Figure 4.14 Composite T-joint finite element model layup detail

Linear softening damage law is adopted to describe the damage evolution. A mixed mode fracture energy criteria is used. Failure under mixed-mode conditions is governed by a power law interaction of the energies required to cause failure in the individual modes. Fracture energy is varied as an input parameter to investigate its effect on crack propagation and make the best match with the measured data.

To characterise the failure behaviour, the distribution and magnitude of the stresses along the joint centreline, starting in the web and progressing round the angle into the join between flange and skin, are extracted as shown in Figure 4.15. In particular, the stresses are plotted in the centre of the coupon and at the surface edge (accessible for observation).

Figure 4.16 and Figure 4.17 show the distribution of the σ_{11} and σ_{12} stresses along the central line of the web (line AB), the interface between the corner and epoxy resin (arc BC) and the interface between the skin and flange (line CD). As can be seen from the graphs, both mode I σ_{11} and mode II σ_{12} stresses have a maximum in the region of the corner, which is where the cracks initiate when the applied load exceeds the critical level. This has been confirmed by the experimental observations. The calculations also show that the mode I stress is always higher than the mode II stress.

Figure 4.18 and Figure 4.19 illustrate the distribution and magnitude of stresses (σ_{11} and σ_{12}) along the interface between the UD-skin and epoxy resin (line OC) and part of line CD (the interface between the skin and flange). The magnitudes of the stresses along the line OC are smaller than those at the corner itself.

The experimental behaviour can be observed by varying the interface fracture toughness as an input to the finite element model. Figure 4.20 shows the effect

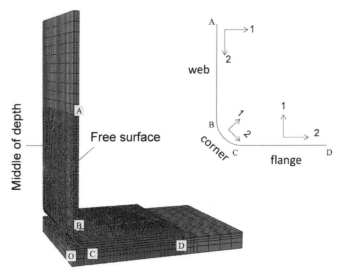

Figure 4.15 Schematic showing the lines for stress interrogation and the directions of stresses σ_{11} and σ_{12}

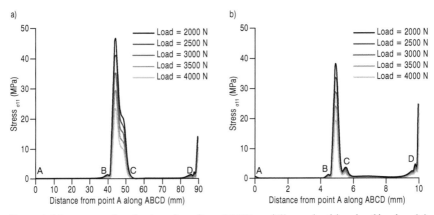

Figure 4.16 σ_{11} stress distribution along line ABCD at different load levels of load at (a) mid-depth and (b) free surface

of fracture energy on the strength to failure; as expected, the ultimate stress to failure increases as the initiation fracture energy increases.

Blade modelling

Parametric blade model

A general finite element modelling tool for wind turbine blades has been implemented in the program Abaqus. The main features of the modelling tool developed are:

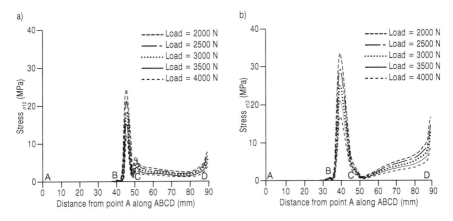

Figure 4.17 σ_{12} stress distribution along line ABCD at different load levels of load at (a) mid-depth and (b) free surface

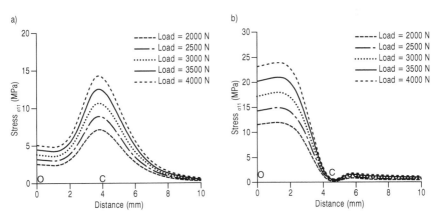

Figure 4.18 σ_{11} stress distribution along line OCD (first part only) at different load levels of load at (a) mid-depth and (b) free surface

- Fully parametric, fully integrated and fully automated blade model
- No need for computer aided design (CAD) geometry
- Smooth surface joining aerofoil sections
- Explicit modelling of the skin/shear-web flange adhesive joints
- Non-linear quasi-static analysis, including main operational loads: gravity, centrifugal and aerodynamic
- Flap and edge bending, torsion, modal and buckling analyses
- Coded for single runs or sensitivity studies on single or multiple parameters.

The blade is modelled in the correct location relative to the ground, turbine tower and hub using axis system transformations to represent the blade forces

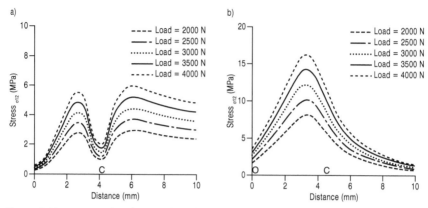

Figure 4.19 σ_{12} stress distribution along line OCD (first part only) at different load levels of load at (a) mid-depth and (b) free surface

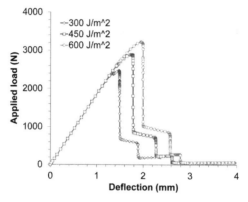

Figure 4.20 T-joint finite element model – effect of varying interface fracture toughness (GIc) on the ultimate stress to failure

and displacements in the local blade axes. This entails the application of the nacelle overhang, but also of any potential rotor tilt, rotor cone and blade pitch angles. The pitch angle and the rotational speed can be specified by the user directly or via a control scheme look-up table. A collective pitching scheme can be selected to specify the blade pitch and rotor speed directly from the wind speed or a cyclic pitching scheme can be selected which takes into account also the azimuth to set the blade pitch angle. More complex control schemes could easily be specified, either through the input of simple look-up tables or through full controller algorithm models incorporated directly into the automation script or accessed via an external program.

A parametric, flexible modelling strategy is sought in order that the blade external geometry, internal structural shape, structural mesh, materials, loads, analysis type and post-processing operations can be specified independently. Working with the commercial FE solver Abaqus (Dassault Systèmes, 2011), these specifications prompted the use of Python scripting (Python, 2009) for model development. All aspects of an analysis can be automated from within the script thanks to the comprehensive features of the Python programming language, augmented by additional functions available to control Abaqus.

As is standard practice in the wind energy field, the aerodynamic shape of the blade is defined by consecutive 2D sections (input via text files of X and Y coordinates) adapted by other inputs such as chord lengths, twist angles, thickness scaling, pitch axis location, etc. The complete aerodynamic surface is then defined by a natural cubic spline interpolation algorithm creating a smooth surface through the successive curves. The numerical generation of the mesh within the script means that neither a CAD package nor a third party mesher is required in the analysis loop. The mesh nodes are then automatically imported into the Abaqus environment and the elements are joined together to form the structural model.

The total control over the meshing algorithm enables a fully structured mesh to be created and the mesh connectivity between the various sub-components is produced automatically. To replace the presence of geometry entities used for the application of materials and boundary conditions, a list of element and node sets is defined. The structured character of the mesh and the data structure in Python largely facilitate these operations. The script also gives the user the choice of using linear (S4R shells and C3D8R bricks) or quadratic (S8R shells and C3D20R bricks) elements. The linear and quadratic shells in Abaqus have slightly different behaviours. The linear S4R element is a general-purpose shell while the quadratic S8R features a thick-shell approximation and a constant thickness.

The blade structural modelling is presently articulated around zero-thickness shell elements for the main composite structural elements (aerodynamic surfaces and internal shear-webs) and 3D brick volumes for the adhesive joints used to connect the composite parts. In the current implementation, the code is restricted to an internal structure composed of two C-type shear-web spars glued to the skins. The possibilities of modelling different types of spar structures and of using ribs will be the subject of future work. The shear webs are defined parametrically by stipulating the distances from root where they start and end as well as their transverse positioning. The four glue layers are inserted between the shear-web flanges and the outer skins, with their thickness also defined as a user parameter in the script. Figure 4.21 shows the layout of the internal structure. The structural characteristics are described through the input of specific data for the material plies – e.g. glass fibre reinforced plastic (GFRP) uni-directional (UD) layer – and layups. Such composites are modelled as orthotropic laminar sections, giving characteristics in the two main directions. A composite layup can then be defined, composed of a succession of plies, each based on a particular material and angular orientation relative to the main directions of the blade. In the layup, each ply can be related to a specific region of the mesh. The whole blade is covered by four layups – one for each outer skin side and one for each shear-web.

Another user parameter provides for the possibility of taking into account non-linear geometric effects. This is necessary when large deflections occur or, for instance, in the case of modelling centrifugal stiffening effects.

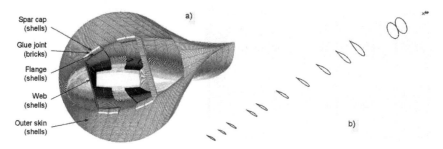

Spar cap
(shells)

Glue joint
(bricks)

Flange
(shells)

Web
(shells)

Outer skin
(shells)

a)

b)

Figure 4.21 Blade model generated by the script and imported into Abaqus: (a) structural representation of the mesh, (b) external surface topology

A sensitivity analysis functionality is built into the script to enable powerful design and scientific studies. This enables the tuning of any parameter used to create the model, including those that affect the blade geometry. This is fully automated as an additional outer loop and does not require user intervention between consecutive runs. For FE analyses, this tool is initially useful for mesh convergence studies. Finally, analysis post-processing is also automated. Key results are output to a text file (e.g. deflections at the blade tip, total mass, sectional mass and stiffness properties, etc.) and Abaqus plots are produced and saved as images. Output data manipulation is also possible.

Aerodynamic loading

One key aspect in realistic structural modelling of wind turbine blades is the accuracy of the loading conditions, which come from several sources. Absolute static strength under extreme loads, fatigue strength under operational loads and tip/tower clearance all bring specific design requirements. As rotor diameter increases, blade mass related requirements become relatively more severe. It is in the interest of the designer to represent the geometric properties underlying these load components parametrically as part of the input script, in order to minimise the time spent in exploring alternative configurations and to reduce the possibility of introducing errors.

Gravity and centrifugal are the most basic loading types to be accounted for. They are nevertheless crucial for a sound representation of blade structural aspects. For large machines, gravity loading is paramount for the edgewise fatigue loads and is also key for problems related to large blade deflections. Centrifugal loading is also important in that it provides a stiffening effect beneficial for tower clearance deflections. Accounting for the centrifugal relief is significant in order to minimise the over-engineering of rotor blades.

For these effects to be modelled accurately, non-linear Abaqus analyses are used. Accurate modelling of gravity and centrifugal loadings is ensured thanks to the finite element discretisation of the structure, accounting for the local materials and layups. Values entered in the input script for the gravity acceleration constant and for the rotational speed of the turbine are used, together with the exact direction of the rotor axis.

Additionally, a fully integrated aerodynamic loading procedure based on a 2D potential flow aerofoil analysis code and the blade element momentum theory is implemented, as described below.

The traditional approach to blade modelling typically involves representing the operational aerodynamic loading as a set of resultant loads at discrete points along the blade, concentrated at the ¼ chord. This is not recommended for 3D FE models, since it has the effect of locking together the degrees of freedom (DOF) of each region where a load is applied, which, in turn, causes two detrimental effects on the model results:

- the local strain/stress field within each loading region is distorted,
- the aerofoil shape may be constrained unrealistically and therefore respond inappropriately.

If wind turbine blade designs are to account for material tailoring of stress/strain patterns and/or for passive or active aerodynamic shape adaptation, modelling tools with realistic aerodynamic load implementation are a key prerequisite.

The application of representative distributed loading on an FE mesh is not a trivial subject. The load should be applied as a distribution on the whole blade surface, either element by element or node by node. Of course, a pressure application element by element is the intuitive choice for wind turbine blades. To apply a full pressure field on the blade, a 2D potential flow panel code, augmented with a viscous boundary layer (BL) analysis, and a blade element and momentum (BEM) algorithm can be used advantageously. These tools are complementary for the study of rotor aerodynamics, and their potential interactions obvious:

- The 2D aerofoil analysis code strictly solves the flow around the profile and outputs the overall aerofoil performance data Cl, Cd and Cm in the form of pressure and friction coefficient distributions around the profile as a function of the angle of attack and Reynolds number;
- The BEM integrates the rotor aerodynamics, solves the wake and induction factors and thereby specifies the aerofoil flow angle of attack and Reynolds number.

The interaction between the two codes ensures a coupled-convergence strategy and enables the aerodynamic pressure to be realistically applied on the whole blade surface.

The first part invoked in this loop is the 2D aerofoil analysis tool. The definition of pressure coefficients is valid for compressible fluids if the fluid density variations are small enough, which can reasonably be assumed for a flow velocity of less than 370 km/h (i.e. < 100 m/s). Provided the aerofoil definition contains enough points (typically > 60), the initial inviscid pressure coefficient (Cp) distribution leads to reasonably accurate lift and pitch moment coefficients.

This solution is presently implemented in two ways in the script and the user chooses which is to be used:

- The model script writes an XFoil command input file with the appropriate aerofoil coordinate file name and angle of attack, runs the XFoil code in batch mode and provides the inviscid analysis results back to the Python script;
- The model script runs an inviscid panel code, fully integrated in the Python script.

XFoil (Drela, 2009) is a purpose-written subsonic aerofoil design and analysis interactive program. It is well regarded by the aerodynamics community for its rigorous implementation, empirical corrections and treatment for low Reynolds numbers. On the other hand, the advantages of the integrated panel code are its simplicity, speed of operation and independence from the use of any third party software. Both implementations were validated successfully against each other and against JavaFoil (Hepperle, 2009), a less well-known user-friendly alternative. Part of these validation results are shown in Figure 4.22, where the pressure coefficient distribution around a Delft University DU97W300 aerofoil at 3° angle of attack is shown for the three implementations. The level of matching is excellent.

However, none of these solutions enable any estimate of the drag. Using an inviscid pressure distribution can therefore be expected to slightly underestimate aerodynamic flap loads and significantly overestimate aerodynamic edge loads, leading to significantly overestimated power predictions. In conventional aerodynamic design work, a boundary layer (BL) analysis typically follows the initial inviscid method to evaluate the friction drag of the aerofoil, related to the boundary layer shear stresses due to the velocity gradient. In XFoil, the solution of this part

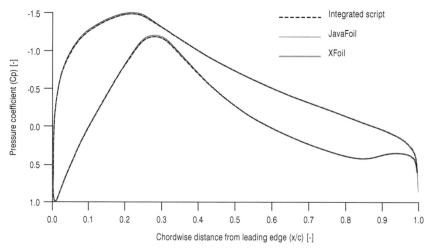

Figure 4.22 Validation of 2D aerofoil pressure distribution code – the Python implementation matches those of XFoil and JavaFoil

is conducted together with an inviscid solution for the flow just outside the BL to ensure convergence as both flows affect each other. At the end of this procedure, XFoil outputs the pressure and friction coefficient distributions along the aerofoil contour to a text file. These distributions can be applied on the FE mesh as pressures and surface tensions, allowing a good representation of lift, drag and pitch moment and hence a significant improvement for the edge load and power estimates. This is the recommended algorithm for use in the script at present and constitutes the current feature under development. This solution also includes separation, transition and stall models and is a good basis for the analysis of pitch-regulated wind turbine blades or of below-rated regime only for stall-regulated machines.

Nevertheless, the treatment of post-stall aerodynamics is mostly empirical and XFoil output should not be relied upon for deep stall applications. To alert the script user, a warning message is issued by the blade script program if the output of the aerodynamic routine takes the aerofoil beyond the maximum lift point (CLmax). This is particularly critical when modelling stall-regulated machines. In such cases, alternative aerodynamic loading routines should be used instead, such as based on experimental measurements and/or on Navier–Stokes solver results, a possible area for future development.

The second main component of the aerodynamic loading calculation is a BEM routine. The code included in the script is very similar to most basic implementations and features the following standard effects and corrections:

- vertical wind shear (logarithmic and exponential models)
- turbulent wake state correction
- tip loss correction
- hub loss correction
- tower shadow (upwind and downwind models).

Industry standard models and conventions (see Freris, 1990; Bossanyi, 2006; Burton et al., 2001) have been adopted for these phenomena. Wind shear is input via a choice of logarithmic or exponential model (Freris, 1990). The logarithmic wind shear model is expressed as:

$$U_z = U_{ref} \frac{\ln \dfrac{z}{z_0}}{\ln \dfrac{z_{ref}}{z_0}}$$

where U_z (m/s) is the undisturbed wind velocity at height z (m), U_{ref} (m/s) is the undisturbed wind velocity at height z_{ref} (m) and z_0 (m) is the surface roughness length. The exponential wind shear model is expressed as:

$$U_z = U_{ref} \left(\frac{z}{z_{ref}} \right)^{\gamma}$$

where $\gamma(-)$ is the wind shear exponent.

Following GH Bladed (Bossanyi, 2006) for heavy rotor loading, a turbulent wake state correction factor H (-) is used whenever the axial induction factor a (-) is greater than 0.3539:

$$H = \frac{4a(1-a)}{0.6 + 0.61a + 0.79a^2}.$$

(Note that the threshold value for a ensures the continuity of the algorithm by enforcing $H = 1$ at $a = 0.3539$.)

The tip and hub losses are modelled using Prandtl's approximation function F (-) applied as a multiplying factor for the axial and tangential inflow factors (Freris, 1990):

$$F = \frac{2}{\pi} arccos\left(\frac{-\frac{B(x-1)}{2}}{e^{\sin(\varphi)}}\right)$$

where x (-) is the ratio R/r when calculating the tip losses and r/R_{root} for the hub losses with R (m) being the rotor radius, r (m) the local radius of interest and R_{root} (m) the radius of the hub, B (-) is the number of blades and φ(rad) the flow incidence angle.

The upwind tower shadow model is based on a potential flow approximation (Bossanyi, 2006):

$$U_{shadow} = U_z\left(1 - \frac{\frac{1}{4}D_{tower}^2 \cdot f_{dia}^2 \cdot (d_{long}^2 - d_{lat}^2)}{(d_{long}^2 + d_{lat}^2)^2}\right)$$

where U_{shadow} (m/s) is the wind velocity at the point in the shadow considered, D_{tower} (m) is the tower diameter at the height considered, f_{dia} (-) is a diameter correction factor provided by the user, d_{long} (m) is the longitudinal distance between the blade section ¼-chord point and the tower centreline and d_{lat} (m) is the corresponding lateral distance.

The downwind model (Bossanyi, 2006) is empirical, based on the concept of a velocity deficit profile that can be shaped proportionally with the distance considered:

$$U_{shadow} = U_z\left(1 - R_{def} \cdot \sqrt{\frac{L_{ref}}{d_{long}}} \cdot \cos^2\left(\frac{\pi d_{lat} \sqrt{\frac{L_{ref}}{d_{long}}}}{D_{ref}}\right)\right)$$

where R_{def} (-) is the maximum velocity deficit ratio in the centreline of the tower wake, D_{ref} (m) is the width of the tower shadow at the corresponding location and L_{ref} (m) is the longitudinal distance from the tower centreline at which R_{def} and D_{ref} are defined. The square root terms act as correcting factors to decrease the velocity deficit and increase the shadow width as the distance from the tower increases. Note that D_{ref} and L_{ref} are input by the user as fractions of the tower diameter for a user-friendlier parameterisation.

The above tower shadow models are applied as specified for azimuth angles within ±60º of the bottom centre position and no tower shadow model is applied for azimuths within ±60º of the top centre position. To ensure a soft transition between these two extreme zones, a smoothing function is applied in the remaining two 60º areas:

$$U_{transition} = U_{shadow}\left(\frac{1}{2}-\cos(\emptyset)\right)+U_z\left(\frac{1}{2}+\cos(\emptyset)\right)$$

where φ(rad) is the azimuth angle.

All these developments were validated against Garrad-Hassan's Bladed software (Bossanyi, 2006) at each stage of their implementation. As an example of this validation, Figure 4.23 shows the angle of attack (AOA) and axial induction factor (AIF) results for the BEM solution of a 2 MW 36 m long blade, taking into account the turbulent wake state correction as well as the wind shear, tower shadow, hub and tip loss models. Figure 4.23 shows that the level of agreement between the two implementations is very high.

It should be noted that each iteration of the BEM wake calculation routine requires one execution of the 2D aerofoil analysis code. In other words, the two processes are dynamically coupled.

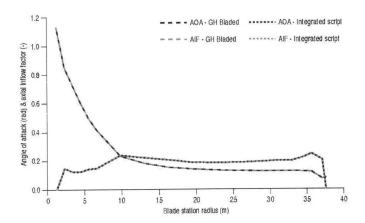

Figure 4.23 Angle of attack and axial induction factor along the rotor radius for the script-integrated BEM code against results from GH Bladed for the exemplar 2 MW design

Once the process has reached a solution to the required accuracy, the pressure coefficients are known around the blade sections concerned. The 2D aerodynamic analysis solver requires a fine aerofoil profile definition that is typically only available at the main blade sections (see Figure 4.21(b)). Typically, the aerodynamic discretisation is finer than the structural analysis one, especially around the areas of larger curvature such as the leading and trailing edges. To apply the pressure coefficient distributions found at these sections to the FE mesh of the whole blade, the first problem to overcome is the profile discretisation discrepancy between the aerodynamic and structural meshes. A marching algorithm, going from the leading edge to the trailing edge on both sides consecutively, is used to sum and apply the pressure coefficients from one mesh to the other. This algorithm does not require any assumption about the relative level of discretisation between the two meshes.

The overall normalised sectional aerodynamic force coefficients C_l, C_d and C_m can be calculated from the pressure coefficients C_p according to the following equations:

$$C_l = \oint C_p d\bar{x}$$

$$C_d = \oint C_p d\bar{y}$$

$$C_m = \oint -C_p \left[\left(x - x_{ref} \right) d\bar{x} + \left(y - y_{ref} \right) d\bar{y} \right]$$

where x_{ref} and y_{ref} are the coordinates of the force concentration point chosen for the pitch moment and where a flow-oriented axis system ($\bar{x}$, $\bar{y}$)s used, according to:

$$\bar{x} = x . \cos(\alpha) + y . \sin(\alpha)$$

$$\bar{y} = y . \cos(\alpha) - x . \sin(\alpha)$$

where α is the local aerofoil angle of attack.

To validate fully the pressure coefficient transfer algorithm, this force calculation was carried out with the C_p series calculated initially on the aerodynamic mesh and then with the C_p series calculated for the FE elements. The level of fidelity of the pressure transcription routine between these two calculations was high, resulting in relative errors on C_l and C_m of around 0.5 per cent, largely independent of the relative aerodynamic and structural mesh sizes. The relative errors for the C_d were of course higher because of the accumulation of numerical errors, especially when more elements are considered, but the resulting forces were still negligibly small compared with the lift forces.

On the finite element mesh, the loads are applied as local element pressures, calculated from the relationship between pressure and pressure coefficient:

$$p = \frac{1}{2}\rho . C_p . V_{local}^2$$

where p is the surface pressure on a finite element, ρ is the air density, C_p is the equivalent pressure coefficient resolved for the finite element considered and V_{local} is the relative velocity of the flow at the blade section considered.

The aerodynamic pressure distributions calculated at the main sections are then processed by a bespoke interpolation routine on all the other mesh sections along the blade main axis, resulting in a full pressure field defined for each finite element of the pressure and suction sides of the blade. In this process, the variations along the blade of the angle of attack, chord length and aerofoil shape are implicitly taken into account via a linear interpolation procedure from one calculated aerofoil station to the next.

The aerodynamic pressure loads are implemented in Abaqus through the use of analytical fields, enabling spatial variations. Equations are input to replicate the linear interpolation functions between the consecutive 2D aerofoil solutions obtained at the discrete number of blade sections. The complete aerodynamic routines and the application of the resulting loads onto the model are fully automated as part of the parametric wind turbine blade model. Figure 4.24 shows a typical resulting pressure field in the graphical user interface.

The developed model is well suited for a first-pass evaluation of potential replacement materials for use in blades. Possible refinements to the model thus far described for more specialist investigation might include:

- corrections to the 2D aerofoil performance coefficients for rotational effects – it is generally recognised that such effects are important at the most inboard sections of wind turbine blades and correction models have been devised for application with the BEM theory (Hansen et al., 2006);
- misalignment between the rotor axis and the wind direction in the BEM model (Hansen et al., 2006), allowing yaw, tilt and cone angles to be taken into account in the aerodynamic loading treatment;
- extension of the aerodynamic loading routines to include dynamic, time-varying effects (e.g. inclusion of aerodynamic damping) to properly model the aero-elastic dynamics of the rotor.

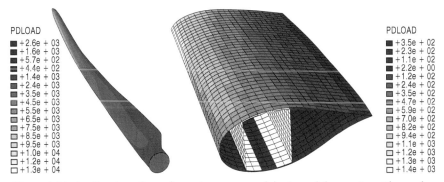

Figure 4.24 Application of aerodynamic pressure – (a) view of the suction side on the 2 MW exemplar turbine blade, (b) close-up on a section near the maximum chord area. Every element has its own pressure application, resulting in a smooth and detailed aerodynamic load treatment that conforms with the overall blade BEM results

Applications of the parametric blade model

To illustrate the main capabilities and features of the modelling tool, input parameter files have been developed for two generic blades intended for a 2 MW onshore turbine and a 5 MW offshore turbine. These blade models are derived from two exemplar designs used more widely within the SUPERGEN Wind consortium. In the results presented below, the gravity, centrifugal and steady state aerodynamic operational loads are implemented; the aerodynamic calculations being conducted with the XFoil solver. Unless otherwise stated, non-linear geometry effects are modelled, so that the centrifugal and large deformation effects can be modelled accurately.

Comparison of the distributed pressure field aerodynamic loading with lumped loading

This case study presents a comparison of the through thickness pressure side spar cap longitudinal stress σ_{11} in the 5 MW blade for analysis runs utilising the distributed aerodynamic load procedure presented above and the more conventional sectional lumped load procedure. This second load application mode was also programmed as an option into the general script, but since, for a yet unidentified reason, the Abaqus solver would not converge for a non-linear analysis with this loading, the comparisons in this section are necessarily conducted as linear analyses.

For the lumped (or concentrated) load application, fictitious ¼-chord points are used to transmit loads rigidly to the local aerofoil section external nodes. In both analyses, the root nodes are rigidly connected to a hub axis point at which the overall force and moment reactions are checked to be the same within 1 per cent.

The through thickness longitudinal stress in the pressure side spar cap for the case computed with the distributed aerodynamic pressures is shown in the 3D plot of Figure 4.25. The stress is plotted against both the laminate thickness and the

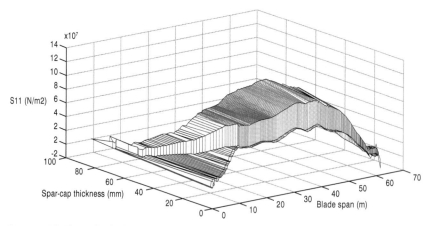

Figure 4.25 Through-thickness longitudinal stress in the 5 MW blade model pressure side spar cap for the distributed aerodynamic load case. The stress variations along the blade span are generally very smooth thanks to the detailed applications of both the layup properties and the loads

blade span. The laminate thickness convention used is such that 0 mm thickness corresponds to the internal side of the composite and the maximum thickness to the external surface of the blade. The spar cap is formed by multi-directional plies – mostly used near the root and which do not transmit the largest stresses – that encapsulate a stack of UD plies. The UD stack takes the highest stresses due to its higher stiffness in the preferential strain direction. Its thickness progressively increases and decreases along the blade span. It can be seen that the external UD plies are slightly more loaded than the internal ones since the strain on the external surface of the laminate is slightly higher. In the blade model used here, the layup of the UD plies in the spar cap is applied with well-distributed ply-drops, ensuring as few stress concentrations as possible. As shown in Figure 4.25, this results in relatively smooth stress variations along the blade span.

The choice of procedure used to apply the aerodynamic loads to the blade mesh is then changed in the script in order to apply the lumped sectional loads at the ¼-chord points. Figure 4.26 presents the same 3D stress plot for this new case.

It can be seen from Figure 4.26 that the stress profiles obtained from the lumped loading are unsurprisingly "polluted" by the sectional aerodynamic load concentrations. Such concentrations are obviously problematic when analyses are conducted to identify the peak local stress in the composite, to investigate specific local design issues, or perform fatigue estimations. Moreover, it should be noted that a non-linear analysis with the lumped loading is less stable for the Abaqus solver than with a distributed pressure load. Overall, it is concluded that the lumped loading procedure is inadequate for many types of blade structural analysis.

Modelling fibre angle induced flap/twist coupling

The careful application of fibre reinforced composite materials has been studied with the intention of providing wind turbine blades with a flap/twist coupling, see

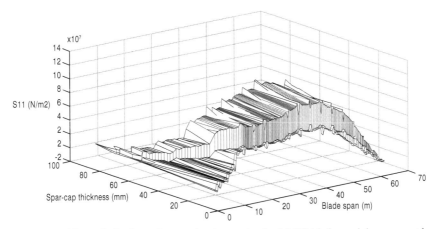

Figure 4.26 Through thickness longitudinal stress in the 5 MW blade model pressure side spar cap with the ¼-chord lumped aerodynamic load case. The stress variations along the blade span are no longer smooth due to the load application

e.g. de Goeij et al. (1999) and Locke and Valencia (2004). For pitch controlled blades, a negative coupling (i.e. positive flap deflections lead to decreasing angles of attack for the same incidence angle) may be a useful feature for load alleviation. In the integrated model developed, changes of material properties affect directly the response of the blade design to operational loads. Moreover, the parametric character of the model enables sensitivity analyses to be run to check what fibre off-axis angle values are most effective. With the SUPERGEN Wind exemplar 2 MW blade design, the fibre orientation angle of the structural glass/polyester UD plies was tuned to values from 0º to –30º relative to the blade longitudinal direction (the negative sign convention implying that the fibres go from the trailing edge towards the leading edge while going from blade root to tip). Azimuth sensitivity analyses were conducted from 0º to 360º in 10º steps and the results are presented in Figure 4.27. Note that the isolated points for each curve on the left-hand side represent the tower passing points (i.e. 180º azimuth) and that the 0º points have been circled for clarity. The exact location of the tower passing points in Figure 4.27 should be ignored since no structural and aerodynamic damping is present in this model. Moreover, it is made clear that no study regarding the dynamic stability of such solutions can be conducted with the present implementation of the script.

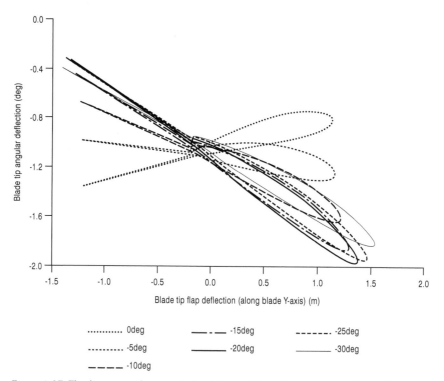

Figure 4.27 Flap/twist coupling graph for different fibre off-axis angle values of the main structural GFRP UD plies. The flap/twist coupling is significantly affected by the choice of this parameter

The first conclusion from Figure 4.27 is that the flap/twist coupling is clearly only one of the structural effects that take place in the blade structure. Indeed, the lack of flap-wise deflection does not result in an unchanged pitch angle for any of the fibre orientations tested – instead, a negative change of around 1º is observed. Moreover, a hysteresis loop characterises all the curves, as opposed to a deterministic flap-wise deflection/tip twist relationship. These phenomena are identified to be mostly due to the torsion moment variations around the pitch axis of the blade from the aerodynamic loading.

Nonetheless, Figure 4.27 shows that, by modifying the fibre direction in this particular selection of plies, the flap/twist coupling behaviour can be significantly affected. The baseline design with no fibre off-axis angle features a positive flap/twist coupling; the twist has a positive contribution to the angle of attack while flap deflections increase. This is shown in Figure 4.27 by a positive average slope of the trend for the 0º curve. However, changing the fibre off-axis angle value from 0º to −15º, the trend is towards a significant decrease in positive coupling and even to the creation and development of a negative coupling. Then, positive flap deflections are coupled with a negative contribution to the angle of attack (the twist varies by around 1º through the azimuth cycle, if ignoring the tower passing extreme), a potentially good behaviour for load reduction of pitch controlled machines – note that the implementation of aerodynamic loading feedback is required before final conclusions can be drawn regarding the exact load reduction gain. Comparatively, little change is observed for off-axis angle values beyond −15º, and even perhaps a reversal of the trend beyond −25º.

Potential benefit of cyclic pitch on blade loads and deflections – static calculations

To demonstrate the usefulness of the model developed, a case study is conducted to evaluate the potential benefits of a cyclic pitch scheme on flap and edge blade root moments and total rotor torque. Blade azimuth sensitivity analyses are produced to study these effects. It should be reiterated that the calculations conducted here are quasi-static. Therefore no aerodynamic damping and vibrations of any origin are accounted for. Moreover, no control dynamics and no control error are introduced in the model. Only qualitative comparisons should be made.

In the baseline run, a collective scheme is implemented through look-up tables relating hub height wind speed with rotor speed and blade pitch for all blades. In this analysis, the hub height wind speed is kept constant in time at 25 m/s, so the collective algorithm simply prescribes fixed rotor speed and pitch angles. This is compared with the output of a cyclic scheme in which the blade pitch angle also depends on the blade azimuth. A zero-mean sinusoidal pitch signal in phase with the azimuth angle is added to the collective value. The peak-to-peak amplitude of this additional signal is set by the user – values from 1º to 5º have been tested. The blade flap and edge root moment and tip twist results are presented in Figure 4.28 and Figure 4.29 respectively.

The sudden variations observed in Figure 4.28 and Figure 4.29 around the azimuth of 180º are due to the tower shadow effects. The amplitudes of these changes are

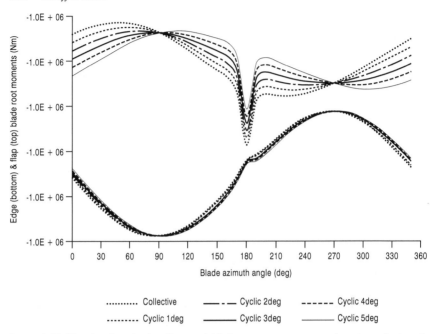

Figure 4.28 Flap (top) and edge (bottom) blade root moments as a function of azimuth angle for different pitch schemes. With the 3° cyclic scheme, the maximum flap load is reduced by 8.4 per cent while the edge load is relatively unchanged

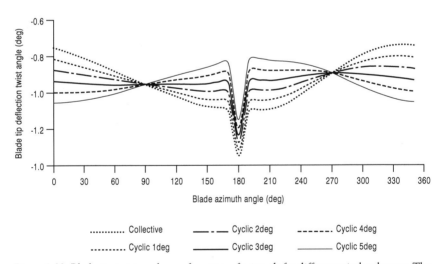

Figure 4.29 Blade tip twist angle as a function of azimuth for different pitch schemes. The tip twist deflection is the most constant with the 3° cyclic scheme

over-estimated in the calculations presented here since no aerodynamic damping or blade dynamics are present. The other main variation that can be seen in these figures is related to the wind shear with a repeating period of 360º.

Figure 4.28 shows that the "cyclic 3 deg" scheme leads to a maximum flap load reduction of 8.4 per cent, without affecting the maximum edge loads. The tip twist angles shown in Figure 4.29 are negative, indicating a negative contribution to the angle of attack. It is seen also that the "cyclic 3 deg" scheme minimises the cyclic variation of blade twist.

In terms of computational efficiency, the computation of each of the six cases shown in Figure 4.28 and Figure 4.29 (i.e. each group of 35 azimuth angle runs) takes about 90 minutes on a 2006 3.4 GHz Intel Pentium D desktop PC for a blade model comprising about 5400 elements and 31,000 DOFs.

Acknowledgements

The materials testing was performed and reported by Chi Zhang, Kuangyi Zhang, Amirhossein Hajdaei, and Zijie, at the Northwest Composites Centre, University of Manchester under the supervision of Professor Paul Hogg and Professor Costas Soutis. The T-joint test coupon finite element modelling was carried out by Dr Ying Wang, also at the Northwest Composites Centre. The parametric 3D blade model was developed initially by Paul Bonnet and later extended by Dr Matthew Clarke and Charles Evans at STFC Rutherford Appleton Laboratory under the supervision of Dr Geoff Dutton. All these contributions to this chapter are gratefully acknowledged.

References

Bossanyi, E., 2006. *GH Bladed version 3.72 User Manual*, Garrad Hassan and Partners Limited.

Burton, T., Sharpe, D., Jenkins, N. & Bossanyi, E., 2001. *Wind Energy Handbook*. Wiley.

Dassault Systèmes, 2011. *Abaqus Analysis User's Manual*. Dassault Systèmes.

de Goeij, W., van Tooren, M. & Beukers, A., 1999. Implementation of bending-torsion coupling in the design of a wind-turbine rotor-blade. *Applied Energy*, Volume 63, pp. 191–207.

Drela, M., 2009. *Xfoil 6.5 user primer*, MIT (see http://web.mit.edu/drela/Public/web/xfoil).

Dulieu-Barton, J.S.P., 1999. Applications of thermoelastic stress analysis to composite materials. *Strain*, 35(2), pp. 41–48.

EWEA, 2011. *UPWIND: Design limits and solutions for very large wind turbines*, European Wind Energy Association.

Freris, L., 1990. *Wind Energy Conversion Systems*. Prentice Hall.

Hansen, M. et al., 2006. State of the art in wind turbine aerodynamics and aeroelasticity. *Progress in Aerospace Sciences*, Volume 4, pp. 285–330.

Harris, B., 2003. *Fatigue in Composites: Science and technology of the fatigue response of fibre-reinforced plastics*. Woodhead Publishing Limited.

Hepperle, M., 2009. *JavaFoil Users Manual*, available from http://www.mh-aerotools.de/airfoils/javafoil.htm (last accessed 11/07/2014).

Janssen, L. et al., 2006. Reliable optimal use of materials for wind turbine rotor blades (OPTIMAT BLADES), Final report, EC project no. NNE5-2001-00174, ECN Wind Energy.

Locke, J. & Valencia, U., 2004. *Design studies for twist-coupled wind turbine blades*, Sandia (SAND2004-0522).

Python, 2009. *Python Reference Manual*. [Online]. Available at: https://docs.python.org/release/3.1/. [Accessed 11/7/2014].

Withers, P., 2007. X-ray nanotomography. *Materials Today*, Volume 10, pp. 26–34.

5 Connection and transmission

*Antony Beddard, Mike Barnes, Antonio Luque,
Alan Ruddell and Olimpo Anaya-Lara*

Offshore turbine array electrical connection and transmission

Seen from the shore, a modern offshore wind farm producing 1000 MW may seem analogous to a conventional power plant. However rather than a few large generators fed by controllable amounts of fossil fuel, the offshore power plant is actually formed by tens or hundreds of individual wind turbines harnessing optimal power from the wind. This power has to be collected by the offshore array of power cables, before being converted to a higher voltage for transmission to shore. For very long distances, it is more economical to use a DC voltage for transmission back to shore. As the offshore collection array voltage is AC, power electronic converter stations in this case would be required to convert the offshore AC to DC for transmission, and then back to AC for connection to the onshore network, as shown in Figure 5.1.

This system is complex and limited information exists in the public domain. The connection to shore, the integration of storage to help manage wind power variation, and the representation of the system in dynamic simulation with appropriate fidelity, is particularly challenging and forms the focus of this chapter. Based on the information contained in this chapter, the development of an example test model of a point-to-point VSC-HVDC link for the connection of a typical Round 3 windfarm is discussed.

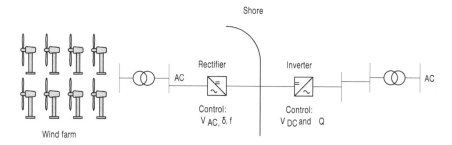

Figure 5.1 Example HVDC connection to shore for an offshore wind farm (single-line diagram)

HVDC connection to shore

Detailed electromagnetic transient models of offshore windfarms, and the connections which facilitate the power transfer back to shore, are extremely important given the role these are likely to play in the future UK energy mix.

The connection for an offshore windfarm can be via either a High Voltage Alternating Current (HVAC) transmission system or a High Voltage Direct Current (HVDC) transmission system. The choice of transmission system is largely dependent upon how far the windfarm is located from shore. Generally speaking HVDC technology is more favourable for windfarms located more than about 80–130 km from shore. This is primarily because in an HVAC system a large proportion of a cable's current-carrying capacity is required to charge and discharge the cable's capacitance every cycle, whereas in an HVDC system, once the cable is charged almost its entire current-carrying capacity is available for real power transfer. At the time of writing the connection distance of AC technology is being pushed to the upper limits of this range, since Voltage Source Converter (VSC)-HVDC is still relatively new and project developers are capitalising the risk associated with this technology during the development phase. As greater experience with VSC-HVDC development is achieved, the break-even point will fall.

Current Source Converters (CSC) and VSC are the two main types of converter technology used in HVDC transmission systems. VSC-HVDC is more suitable for offshore windfarms because it does not require a strong AC system and it has a smaller footprint in comparison to CSC-HVDC.

The UK Transmission System Operator, National Grid, has presented three different strategies for the connection of the UK's Round 3 offshore windfarms in their Offshore Development Information Statement (ODIS) (National Grid, 2011). The reader should be aware that ODIS has now been replaced with National Grid's Electricity Ten Year Statement. These three strategies have different hardware requirements. For instance the radial strategy requires the equivalent of more than twenty-five 1000 MW VSC-HVDC point-to-point schemes. According to ODIS, the DC voltage for such a link is ±300 kV and the average HVDC cable length of suitable Round 3 sites is approximately 165 km.

The key component of the VSC-HVDC transmission scheme is the converter. Since its inception in 1997, and until 2010, all VSC-HVDC schemes used two or three level VSCs (Barnes, 2012). In 2010, the Trans Bay cable project became the first VSC-HVDC scheme to use Modular Multi-level Converter (MMC) technology. The MMC has numerous benefits in comparison to two or three level VSCs; chief among these is reduced losses. Today, the main HVDC manufacturers offer a VSC-HVDC product which is based on multi-level converter technology; hence this is the focus of the considerable research in this field.

MMC VSC-HVDC

There are several types of MMC, including half-bridge (HB), full-bridge and alternate arm converter (Merlin *et al.*, 2014). The HB is the only type of MMC which is commercially in operation and is therefore the focus of this section.

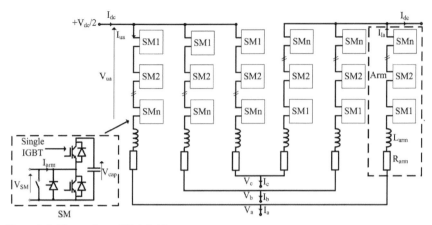

Figure 5.2 Three-phase HB-MMC

The basic structure of a three-phase HB-MMC is shown in Figure 5.2. Each leg of the HB-MMC consists of two converter arms which contain a number of sub-modules (SMs), and a reactor, L_{arm}, connected in series. Each SM contains a two-level HB converter with two IGBTs and a parallel capacitor. The SM is also equipped with a bypass switch to remove the module from the circuit in case an IGBT fails, and a thyristor, to protect the lower diode from overcurrent in the event of a DC side fault.

The converter is placed into energisation mode when it is first powered-up. In order to energise the SM capacitor, the upper and lower IGBTs are switched-off. If the arm current is positive then the capacitor charges and if the arm current is negative then the capacitor is bypassed. This energisation state does not occur under normal operation.

The SM terminal voltage, V_{SM}, is effectively equal to the SM capacitor voltage, V_{cap}, when the upper IGBT is switched-on and the lower IGBT is switched-off. The capacitor will charge or discharge based upon the arm current direction. With the upper IGBT switched-off and the lower IGBT switched-on, the SM capacitor is bypassed and therefore V_{SM} is effectively at zero volts. Each arm in the converter acts like a controllable voltage source, with the smallest voltage change being equal to the SM capacitor voltage. With reference to Figure 5.2, the following equations for the phase A converter voltage can be derived (Antonopoulos, 2009):

$$V_a = \frac{V_{dc}}{2} - V_{ua} - L_{arm}\frac{dI_{ua}}{dt} - R_{arm}I_{ua}$$

$$V_a = V_{la} - \frac{V_{dc}}{2} + L_{arm}\frac{dI_{la}}{dt} + R_{arm}I_{la}$$

The upper and lower converter arm currents, I_{ua} and I_{la}, consist of three main components (Qingrui, 2010):

$$I_{ua} = \frac{I_{dc}}{3} + \frac{I_a}{2} + I_{circ}$$

$$I_{la} = \frac{I_{dc}}{3} - \frac{I_a}{2} + I_{circ}$$

The current component which is common to both arms, $((I_{dc}/3) + I_{circ})$, is commonly referred to as the difference current, I_{diff}. The circulating current, I_{circ}, is due to the unequal DC voltages produced by the three converter legs. Substituting terms and rearranging the above equations gives:

$$V_a = V_{ca} - \frac{L_{arm}}{2} pI_a - \frac{R_{arm}}{2} I_a$$

where p indicates differentiation and the internal converter voltage, V_{ca}, is given by:

$$V_{ca} = \frac{V_{la} - V_{ua}}{2}$$

This shows that the converter phase voltages are effectively controlled by varying the lower and upper arm voltages, V_{la} and V_{ua}. Each converter arm contains a number of SMs, n. Assuming that the SM capacitance is large enough to neglect ripple voltage and that the capacitor voltages are well balanced, the SM capacitor voltage, V_{cap}, can be described by:

$$V_{cap} = \frac{V_{dc}}{n}$$

The voltage produced by an arm of the converter is equal to the number of SMs in the arm which are turned-on, n_{onua} and n_{onla}, multiplied by the SM capacitor voltage, as given by:

$$V_{ua} = n_{onua} \times V_{cap}$$

$$V_{la} = n_{onla} \times V_{cap}$$

The output voltage magnitude and phase can be controlled independently via appropriate control of the SMs. The number of voltage levels that an MMC can produce at its output is equal to the number of SMs in a single arm plus one.

Control systems for an MMC connected to an active network

For a VSC-HVDC scheme which connects two active networks, one converter controls active power (or contributes to AC frequency control at its end) and the

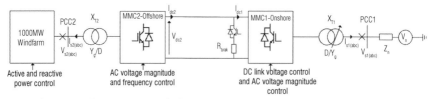

Figure 5.3 Example VSC-HVDC link for the connection of an offshore windfarm

Figure 5.4 MMC control system basic overview

other converter controls the DC link voltage. The converters at each end of the link are capable of each controlling reactive power or the AC voltage magnitude at their respective Point of Common Coupling (PCC). For VSC-HVDC links which are employed for the connection of offshore windfarms, the converter connected to the onshore AC grid controls the DC link voltage and the reactive power or AC voltage at the PCC as shown in Figure 5.3.

Active power, frequency and DC link voltage are effectively controlled by varying the angle of the MMC output voltage with respect to the voltage angle of the connected AC network. The reactive power and AC voltage are effectively controlled by varying the magnitude of the MMC output voltage with respect to that of the AC network.

Figure 5.4 shows the basic MMC control structure used for the example test model developed in this chapter.

The current controller is typically a fast feedback controller, which produces a voltage reference for the converter based upon the current references from the outer feedback controllers. In this work a positive sequence dq (direct and quadrature) current controller is used because it can limit the phase currents under balanced operating conditions and provide a faster response than direct control of the voltage magnitude and phase. More complex controllers such as negative sequence dq control and proportional resonance control can also be used to limit currents under unbalanced conditions. The inner MMC control system, amongst other functions, translates the voltage set-points into Firing Signals (FS) for the MMC SMs to obtain the desired output voltage magnitude and phase.

dq current controller

The impedance between the internal converter voltage, V_{ca}, and the AC system voltage, V_{sa}, for phase A is shown in Figure 5.5. Z_n, R_T and L_T denote the network impedance, transformer resistance and transformer inductance respectively.

The relationship between the internal converter voltage and the AC system voltage for phase A is given by:

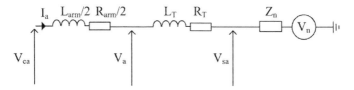

Figure 5.5 MMC phase A connection to AC system

$$V_{ca} - V_{sa} = \left(\frac{L_{arm}}{2} + L_T \right) \frac{dI_a}{dt} + \left(\frac{R_{arm}}{2} + R_T \right) I_a$$

This in turn can be reduced to:

$$V_{csa} = L \frac{dI_a}{dt} + RI_a$$

where:

$$V_{csa} = V_{ca} - V_{sa} \qquad L = \frac{L_{arm}}{2} + L_T \qquad R = \frac{R_{arm}}{2} + R_T$$

For the three-phases, where $p = d/dt$:

$$\begin{pmatrix} V_{csa} \\ V_{csb} \\ V_{csc} \end{pmatrix} = (R + pL) \begin{pmatrix} I_a \\ I_b \\ I_c \end{pmatrix}$$

Applying the abc to αβ (Clarke) transform:

$$\begin{pmatrix} V_\alpha \\ V_\beta \\ V_0 \end{pmatrix} = \frac{2}{3} \begin{pmatrix} 1 & -\frac{1}{2} & -\frac{1}{2} \\ 0 & \frac{\sqrt{3}}{2} & -\frac{\sqrt{3}}{2} \\ \frac{1}{2} & \frac{1}{2} & \frac{1}{2} \end{pmatrix} \begin{pmatrix} V_a \\ V_b \\ V_c \end{pmatrix}$$

gives:

$$\begin{pmatrix} V_\alpha \\ V_\beta \\ V_0 \end{pmatrix} = (R + pL) \begin{pmatrix} I_\alpha \\ I_\beta \\ I_0 \end{pmatrix}$$

Providing that the three-phase system (abc) is balanced, the three AC quantities are transformed into two ac quantities of equal magnitude separated by

90°. The two ac quantities can then be transformed into two dc quantities using the αβ to dq (Park) transform:

$$\begin{bmatrix} V_d \\ V_q \end{bmatrix} = \begin{bmatrix} \cos\theta & \sin\theta \\ -\sin\theta & \cos\theta \end{bmatrix} \begin{bmatrix} V_\alpha \\ V_\beta \end{bmatrix}$$

The Park transform in vector form is given by:

$$V_{dq} = V_{\alpha\beta} e^{-j\theta}$$

where its inverse is:

$$V_{\alpha\beta} = V_{dq} e^{j\theta}$$

The plant equation can thus be re-written in vector form as:

$$V_{dq} e^{j\theta} = R I_{dq} e^{j\theta} + Lp\left(I_{dq} e^{j\theta}\right)$$

Using partial differentiation and noting that $\theta = \omega t$ gives:

$$p\left(I_{dq} e^{j\theta}\right) = e^{j\theta} \left.\frac{\partial}{\partial t}\right|_{\theta=const} \left(I_{dq}\right) + I_{dq} \left.\frac{\partial}{\partial t}\right|_{I=const} e^{j\theta}$$

$$= e^{j\theta} p I_{dq} + j\omega I_{dq} e^{j\theta}$$

Applying this to the plant equation gives:

$$V_{dq} e^{j\theta} = R I_{dq} e^{j\theta} + L\left(e^{j\theta} p I_{dq} + j\omega I_{dq} e^{j\theta}\right)$$

which can be simplified as:

$$V_{dq} = R I_{dq} + \left(Lp + j\omega L\right) I_{dq}$$

or in matrix form:

$$\begin{bmatrix} V_d \\ V_q \end{bmatrix} = R \begin{bmatrix} I_d \\ I_q \end{bmatrix} + Lp \begin{bmatrix} I_d \\ I_q \end{bmatrix} + \omega L \begin{bmatrix} 0 & -1 \\ 1 & 0 \end{bmatrix} \begin{bmatrix} I_d \\ I_q \end{bmatrix}$$

where $\begin{bmatrix} 0 & -1 \\ 1 & 0 \end{bmatrix}$ is the matrix representation of the imaginary unit j. Expanding

and noting that $V_d = V_{cd} - V_{sd}$ and $V_q = V_{cq} - V_{sq}$ gives:

$$V_{cd} - V_{sd} = R I_d + Lp I_d - \omega L I_q$$

$$V_{cq} - V_{sq} = R I_q + Lp I_q + \omega L I_d$$

The equivalent circuit diagrams for the plant in the dq reference frame are given in Figure 5.6.

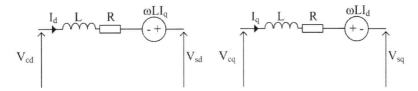

Figure 5.6 Equivalent dq circuit diagrams

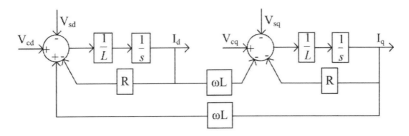

Figure 5.7 SFSB diagram for system plant in dq reference frame

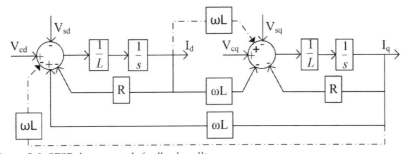

Figure 5.8 SFSB diagram with feedback nulling

Applying the Laplace transform with zero initial conditions gives:

$$V_{cd}(s) - V_{sd}(s) = RI_d(s) + LsI_d(s) - \omega LI_q(s)$$

$$V_{cq}(s) - V_{sq}(s) = RI_q(s) + LsI_q(s) + \omega LI_d(s)$$

The plant equations in the Laplace domain can be represented by the State-Feedback System Block (SFSB) diagrams with the (s) notation neglected, as shown in Figure 5.7.

The SFSB diagram in Figure 5.7 clearly shows that there is cross-coupling between the d and q components. The effect of the cross-coupling can be reduced by introducing feedback nulling, which effectively decouples the d and q components as shown in Figure 5.8.

The d and q currents are controlled using a feedback PI controller as shown in Figure 5.9. The d and q components of the system voltage (V_{sd} and V_{sq}) act

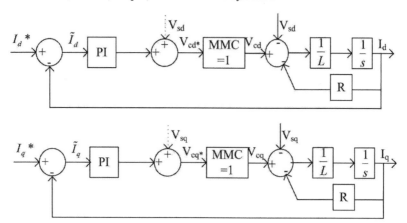

Figure 5.9 Decoupled d and q current control loops

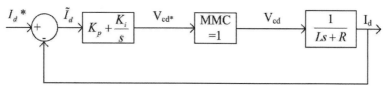

Figure 5.10 d-axis current loop without d-axis system voltage disturbance

as a disturbance to the controller. The effect of this disturbance is mitigated through the use of feed-forward nulling, highlighted in Figure 5.9. The MMC is represented as a unity gain block (i.e. $V_{cd}^* = V_{cd}$), which is representative of its operation providing that the converter has a high level of accuracy with a significantly higher bandwidth than the current controller. The d-axis current control loop in Figure 5.9 can be simplified to Figure 5.10, due to the cancellation of the disturbance term and the use of Mason's rule. This is equally applicable to the q-axis.

The transfer function for the control loop can be calculated as follows:

$$\frac{I_d}{I_d^*} = \frac{\frac{1}{s}(K_i + K_p s)\left(\frac{1}{Ls+R}\right)}{1 + \frac{1}{s}(K_i + K_p s)\left(\frac{1}{Ls+R}\right) \times 1}$$

$$\Rightarrow \frac{\frac{K_i + K_p s}{L}}{s^2 + \left(\frac{R+K_p}{L}\right)s + \frac{K_i}{L}}$$

Approximating the current loop transfer function to a classic 2nd order transfer function:

$$\frac{\omega_n^{\,2}}{s^2 + 2\omega_n \zeta s + \omega_n^{\,2}}$$

$$\Rightarrow \frac{\dfrac{K_i}{L}}{s^2 + \left(\dfrac{R + K_p}{L}\right)s + \dfrac{K_i}{L}}$$

This allows the PI controller to be tuned to give an approximate damping ratio, ζ, and natural frequency, ω_n, using:

$$\omega_n = \sqrt{\frac{K_i}{L}}$$

$$\zeta = \frac{R + K_p}{2\omega_n L}$$

The natural frequency of the system is approximate to its bandwidth for a damping ratio of 0.7 to 1.0. The closed loop transfer function can also be reduced to a first order transfer function if desired.

The resultant block diagram is shown in Figure 5.11. The d-axis and q-axis current orders from the outer controller have limits to prevent valve over-current under balanced conditions.

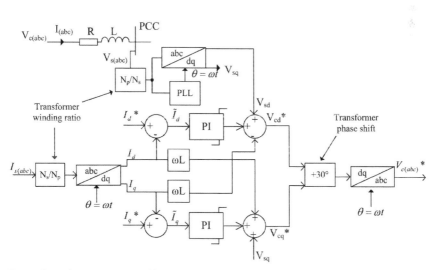

Figure 5.11 dq current controller implementation

Outer controller power loops

In the magnitude invariant dq synchronous reference frame, the power flow at the PCC can be described by:

$$S_{dq} = \frac{3}{2}V_{dq}I_{dq}^* = \frac{3}{2}(V_{sd} + jV_{sq})(I_d - jI_q)$$

$$P = \frac{3}{2}(V_{sd}I_d + V_{sq}I_q)$$

$$Q = \frac{3}{2}(V_{sq}I_d - V_{sd}I_q)$$

Aligning the q-axis with V_a such that $V_{sq} = 0$, allows these equations to be reduced to:

$$P = \frac{3}{2}V_{sd}I_d$$

$$Q = -\frac{3}{2}V_{sd}I_q$$

Outer loop DC link voltage controller

MMCs, unlike two-level VSCs, do not normally employ DC side capacitor banks and therefore the MMC's equivalent capacitance, $C_{eq,}$ is normally employed for the plant model:

$$\frac{3C_{SM}}{n} \le C_{eq} \le \frac{6C_{SM}}{n}$$

The lower limit of C_{eq} is based on the MMC's in-circuit capacitance under normal operation while the upper limit is based on the MMC's total stored energy. The upper limit is often employed as it tends to give a more accurate representation of the MMC's overall DC voltage dynamics.

With reference to Figure 5.12 the DC link voltage can be described by:

$$C_{eq}\frac{dV_{dc}}{dt} = I_n + I_{dc}$$

Assuming negligible converter losses, the power balance between the AC and DC system can be described by:

$$I_{dc}V_{dc} = 1.5V_{sd}I_{sd}$$

Hence:

$$\frac{dV_{dc}}{dt} = \frac{I_n}{C_{eq}} + \frac{3V_{sd}I_{sd}}{2C_{eq}V_{dc}}$$

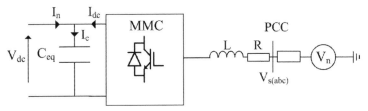

Figure 5.12 DC side plant

Taking partial derivatives gives:

$$\frac{d\Delta V_{dc}}{dt} = \frac{1}{C_{eq}}\Delta I_n - \frac{3V_{sdo}I_{sdo}}{2C_{eq}V_{dco}^2}\Delta V_{dc} + \frac{3V_{sdo}}{2C_{eq}V_{dco}}\Delta I_{sd}$$

$$C_{eq}\frac{d\Delta V_{dc}}{dt} = \Delta I_n - 1.5K_V K_G \Delta V_{dc} + 1.5K_V \Delta I_{sd}$$

where the subscript 'o' denotes operating point and:

$$K_V = \frac{V_{sdo}}{V_{dco}} \qquad\qquad K_G = \frac{I_{sdo}}{V_{dco}}$$

The SFSB block diagram for the DC voltage control loop is shown in Figure 5.13.

The transfer function for the control loop can be derived as follows:

$$\frac{\Delta V_{dc}}{\Delta V_{dc}{}^*} = \frac{1.5K_V(sK_p + K_i)}{C_{eq}s^2 + 1.5K_V(K_G + K_p)s + 1.5K_v K_i}$$

Typically $K_G << K_p$ and hence:

$$\frac{\Delta V_{dc}}{\Delta V_{dc}{}^*} = \frac{1.5K_V(sK_p + K_i)}{C_{eq}s^2 + (1.5K_V K_p)s + 1.5K_v K_i}$$

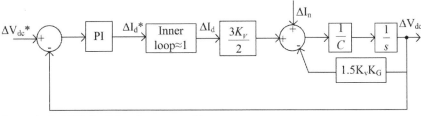

Figure 5.13 SFSB for DC voltage control loop

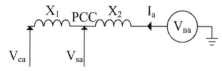

Figure 5.14 MMC phase A connection to the AC network with the system resistances neglected

Rearranging and ignoring the sK_p term gives:

$$\frac{\Delta V_{dc}}{\Delta V_{dc}{}^{*}} = \frac{1.5 K_V K_i / C_{eq}}{s^2 + \left(1.5 K_V K_p / C_{eq}\right)s + 1.5 K_V K_i / C_{eq}}$$

This is a second order approximation from which the natural frequency and damping ratio can be found. The primary function of the DC link voltage controller is to keep the DC link voltage constant. The outer DC link voltage loop is tuned assuming that the inner current loop is a unity gain block. This is a valid assumption providing that the outer loop is significantly slower than the inner current loop.

Outer loop AC voltage control

It may be desirable to control the onshore network voltage, especially for systems where the onshore converter is large compared with the rest of the system and the converter thus has a significant 'grid forming' function.

With reference to Figure 5.14, the magnitude of the voltage at the PCC is given by:

$$V_{sa} = \sqrt{\left(V_{na}(1-k) + V_{ca}k\cos(\delta_c)\right)^2 + \left(V_{ca}k\sin(\delta_c)\right)^2}$$

where:

$$k = \frac{X_2}{X_1 + X_2}$$

The network voltage, V_{na}, and the system reactance values, X_1 and X_2, are fixed and therefore the voltage at the PCC, V_{sa}, can be controlled by varying the internal converter voltage magnitude, V_{ca}, and angle, δ_c, where the network voltage is taken as the reference ($\delta_n = 0$). The dominant variable is the converter voltage magnitude.

Supplementary MMC internal control

Additional controllers are required for the complex power electronic structure which forms the SM chains. Here a brief review of the main controllers is provided.

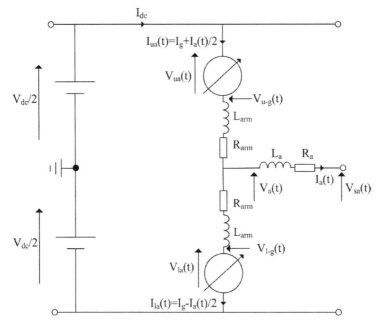

Figure 5.15 Equivalent circuit for a single phase of an MMC

Circulating Current Suppressing Control (CCSC)

Currents circulate between phase legs in the MMC due to voltage imbalance between the phase leg voltages. The circulating currents distort the arm currents, which can increase converter losses and may result in the requirement for higher rated components. The CCSC suppresses the circulating current by controlling the voltage across the limb reactors. The development of the controller shown here is based on work carried out by Qingrui (2010).

With reference to Figure 5.15, the plant equations for the CCSC can be derived as follows.

$$V_{dc} = V_{ua} + I_{ua}(R + Lp) + I_{la}(R + Lp) + V_{la}$$

The voltage drop across the upper and lower arm impedances, due to the $I_a/2$ current components in the upper and lower arm, cancel each other out. I_{ut} and I_{la} can therefore be replaced with $I_{diff} = I_g + I_{circ}$ where $I_g = I_{dc}/3$, hence:

$$V_{dc} = V_{ua} + V_{la} + 2I_{diff}(R_{arm} + L_{arm}p)$$

$$\frac{V_{dc}}{2} - \frac{V_{ua} + V_{la}}{2} = I_{diff}(R_{arm} + L_{arm}p)$$

The right-hand side of this equation is known as the difference voltage, V_{diff}, which is the voltage drop across one converter arm due to the difference current. If the circulating current is zero then the difference voltage is essentially very small as it is the voltage drop across the arm resistance due to the DC current in the arm, I_g. The presence of circulating current increases the difference voltage, hence by reducing the difference voltage, the circulating current can be suppressed.

The difference voltage can be controlled by varying the upper and lower arm voltages. The circulating current is a negative sequence (a-c-b) current at double the fundamental frequency. The plant equation in matrix form is:

$$\begin{bmatrix} V_{diff-a} \\ V_{diff-c} \\ V_{diff-b} \end{bmatrix} = R \begin{bmatrix} I_{diff-a} \\ I_{diff-c} \\ I_{diff-b} \end{bmatrix} + Lp \begin{bmatrix} I_{diff-a} \\ I_{diff-c} \\ I_{diff-b} \end{bmatrix}$$

Applying the acb to dq transform gives:

$$\begin{bmatrix} V_{diff-d} \\ V_{diff-q} \end{bmatrix} = R_{arm} \begin{bmatrix} I_{circ-d} \\ I_{circ-q} \end{bmatrix} + L_{arm} p \begin{bmatrix} I_{circ-d} \\ I_{circ-q} \end{bmatrix} + 2\omega L_{arm} \begin{bmatrix} 0 & -1 \\ 1 & 0 \end{bmatrix} \begin{bmatrix} I_{circ-d} \\ I_{circ-q} \end{bmatrix}$$

The I_{diff} component has changed to I_{circ}, because the DC component of I_{diff} is a zero sequence component which has no effect on the dq values. The dq components are therefore only affected by the circulating current. The SFSB with feedback decoupling is shown in Figure 5.16.

The implementation of the controller is shown in Figure 5.17. The set-point for the CCSC is zero in order to reduce the circulating current to the smallest possible value. The bandwidth of the controller has little effect on the circulating current under steady-state conditions. A small controller bandwidth such as 10 Hz is therefore suitable; however higher bandwidths can provide better performance during transient conditions.

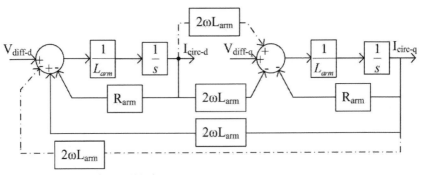

Figure 5.16 CCSC plant SFSB diagram

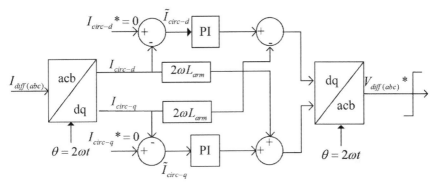

Figure 5.17 Block diagram of CCSC implementation

MMC voltage output synthesis

The MMC voltage output synthesis controller block is required to produce an output voltage from a combination of SM voltages. It determines the number of SMs to fire to produce the voltage reference for each arm, and which SMs to fire to ensure that the capacitor voltages are balanced.

The capacitor balancing controller (CBC) ensures that the energy variation in each converter arm is shared equally between the SMs within that arm. Without such a control system the SM capacitors would exceed their tolerable voltage limits, which could damage the SM IGBTs.

The CBC can be based on the method outlined by Marquardt (2002). The CBC samples the SM capacitor voltages and then sorts them into ascending or descending order based on the direction of the arm current. If the arm current is positive then the SMs with the lowest capacitor voltages are placed first. Conversely, if the arm current is negative the CBC orders the SMs with the highest capacitor voltage first. This ensures that for positive arm current the capacitors with the lowest voltages are charged first and for negative arm current the capacitors with the highest voltages are discharged first.

The other function of MMC voltage synthesis is the switching instant selection algorithm. A number of modulation methods have been proposed for MMCs. These include, but are not limited to, space vector modulation (Lesnicar, 2003), phase-disposition modulation (Saeedifard, 2010), selective harmonic elimination (Zhong, 2006) and nearest level control (NLC) (Qingrui, 2011). The NLC method produces waveforms with an acceptable harmonic content with a suitable number of levels and it is the least computationally complex method of the aforementioned techniques.

The NLC reads in the reference voltage for the upper and lower arm of each phase. Using the DC voltage measurement, it calculates the average SM capacitor voltage and then calculates the exact number of SM levels (ENL) required in circuit to obtain the reference voltage. Due to the finite number of SMs the ENL is rounded to the nearest level. The NLC then issues firing signals (FS) to the correct SMs using the ordering information from the CBC.

Control systems for MMC connected to a windfarm

Publications on the control of MMCs connected to windfarms are sparser than for MMCs connected to active networks. In Mohammadi (2013) the VSCs connected to offshore windfarms were configured to control the frequency and voltage magnitude of the offshore AC networks. The AC voltage and frequency for an offshore network can be controlled with or without an inner current loop (Gnanarathna 2010; Jun 2009). The offshore MMC employed for the example test model controls the offshore AC voltage magnitude and frequency without an inner current loop. This method was found to offer good stability, however the arm currents cannot be limited for an offshore AC network fault without additional control. The internal MMC controls employed for the offshore MMC are the same as for the onshore MMC.

The offshore array

In an offshore wind farm, the purpose of the collection network is to gather together the power outputs of all the turbines in the wind farm at one point for transmission to shore. At the time of writing, the wind turbines used offshore are largely based on designs for onshore use, producing an AC output for direct connection to the electricity grid, and complying with the relevant grid codes for power quality and fault response. For a small wind farm, close to shore, the outputs of the turbines are collected and stepped up using a transformer on a transmission platform, with an AC link to shore. Therefore the collection grid represents an extension of the national electricity grid.

Many planned offshore wind farms will be of a large size, of up to several GW, and will be located a long distance from the shore. To transmit power over long distances undersea, high voltage DC transmission (HVDC) is required, using converter stations on an offshore transmission platform and onshore at the point of connection to the national electricity grid. In this case, the collection network is not a part of the national electricity grid, and it is the responsibility of the onshore converter station to comply with the grid codes, with (possibly) assistance from the wind farm hardware. This allows a much greater range of configurations for the collection network, which could have the following advantages:

- Decreased cost, through the reduction in the number of power electronic converter steps in a DC collection grid, or the use of cheaper DC cables;
- Decreased losses, through the reduction in the number of power electronic converter steps with a DC system, or the use of high frequency transformers;
- Decreased operation and maintenance costs, and increased availability, by having the turbine power electronic converters together on a platform.

Overview of collection network architectures

Several potential collection network architectures have been identified, and many of these have their own variations. The network architectures can be divided into those which use a converter on the turbine, and those where the converter is on a separate platform in order to allow easier access for maintenance. The latter can also be divided into architectures featuring a single converter for multiple turbines, and architectures where each turbine has its own converter. In all systems, it is assumed that the wind farm is located a sufficient distance offshore that HVDC transmission will be required. Here the principal differentiator is how the power electronic conversion of the turbine generator output is handled, i.e. whether:

- power converters are located on the turbine, or
- power converters are located on the collector platform.

Converters on turbine

AC string

The conventional arrangement is shown in Figure 5.18. Turbines feature a squirrel cage induction generator (SCIG), or a permanent magnet generator (PMG) connected to a fully-rated converter, or alternatively a doubly-fed induction generator (DFIG) and partially rated converter can be used. The output of the converter is stepped up to the collection network voltage, and the turbines are connected together in strings.

The number of turbines on a string is determined by the current and voltage rating of the available cables, and the rated power of the turbines. Voltage is limited by the water-treeing effect with wet insulation cables, while dry insulation cables with a

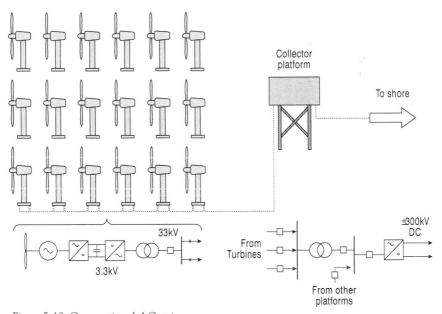

Figure 5.18 Conventional AC strings

lead sheath around the insulation would be too expensive (Crine, 1998). A higher voltage also requires a higher voltage rating for the transformer, which increases the cost and size. Available current ratings are also limited, as the skin depth of the AC current means that conductors with larger areas will be less effective, as the current will not flow in the centre of the cable. For this reason, the cost of AC cables tends to increase exponentially with current capacity (Lundberg, 2003).

While the diagram shows strings of turbines which are connected at one end, the ends of the strings can also be connected together, which allows the end turbines in a string to still export power if a fault occurs in one of the cables in the string. If the turbines are arranged in such a loop, the loop can either be rated such that each connection to the platform can carry the full loop power, giving maximum redundancy, or half power, leading to a loss of power at higher wind conditions (Sannio, 2006). In most cases, the cable reliability is considered to be high, and a loop system is not used. In this case, the cable rating is tapered along the length of the cable to save cost.

DC string

An arrangement using DC in the collection strings is shown in Figure 5.19. In this system, the turbines output a DC voltage, which is then stepped up to the transmission voltage at the collection platform. In most studies, the turbines produce a voltage of around 40–50 kV DC, which requires an AC-DC converter capable of producing such a voltage, featuring many switching devices in series, or a lower voltage AC-DC converter and step-up DC-DC converter (Lundberg, 2003;

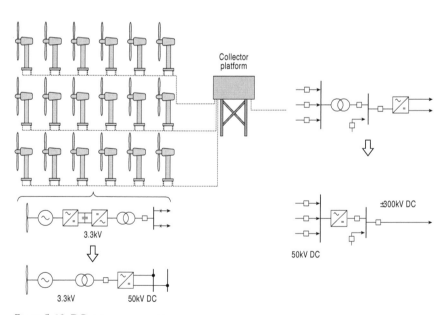

Figure 5.19 DC strings connection

Zhan, 2010). A solution involving a lower voltage converter, and a DC voltage of 5 kV is also possible, which has the advantage of eliminating the turbine transformer and using a conventional 3.3 kV 3-level converter (Monjean, 2010). However, the currents in the strings will be extremely high, requiring thick cable and leading to high losses. DC systems are attractive as they could reduce the number of conversion steps between AC and DC, but converters with a high voltage boost ratio will require a transformer, requiring conversion to AC and back.

As DC cables do not suffer from water-treeing degradation, higher voltages could be used without needing dry-insulation cables (Crine, 1998), while the current in a DC cable can use the entire area of the conductor, so the cable cost will increase linearly with current capacity rather than exponentially as with AC (Lundberg, 2003). Because of these factors, it could be possible to implement longer turbine strings much more cheaply with DC than with AC collection. However, this is difficult to quantify as there are no commercially available cables with the required configuration and voltage rating, and previous studies of the cable cost have extrapolated the cost for multi-core DC collection cables from the costs for single-core HVDC transmission cables with a significantly higher voltage rating.

Another issue with DC collection networks is with fault protection, as the fact that the current does not continually reverse as with AC means that when a circuit breaker opens, the switching arc will not be automatically extinguished when the current reverses (Yang, 2010). Various DC circuit breaker designs have been proposed, but these become increasingly expensive at higher voltage ratings (Meyer, 2005). DC collection and transmission networks have been designed which use power electronic converters which are capable of stepping down the voltage as well as stepping up, and these can be used to limit the fault current, but at the cost of extra complexity (Zhan, 2010).

DC series

An alternative DC collection architecture is to use series DC connection of the turbines, shown in Figure 5.20. Here the DC outputs of the turbines are connected in series, and the turbines are connected in a loop. This allows the high collection voltage to be achieved without using high voltage converters, although the converter would need to be isolated with respect to ground. An isolation transformer would need to be used, or a generator capable of handling a high voltage offset. Another option is to use a transformer isolated converter in the turbine, where the high voltage side of the converter only consists of a passive diode rectifier, which is much easier to isolate (Parasai, 2008).

This arrangement could reduce the cable costs, as it only uses a single core cable loop, although there is no scope to taper the current rating of the cable. In the event of a turbine fault, the faulty turbine could be bypassed using a mechanical switch, but any cable faults will mean that none of the turbines on the loop would be able to export power.

A related idea is to increase the turbine output voltage and the length of the strings, so that the full transmission voltage is produced, eliminating the need for

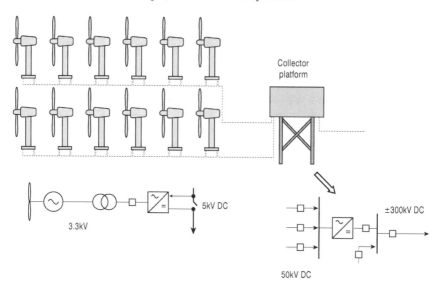

Figure 5.20 Series DC connection

the collection platform (Parasai, 2008). This system has been shown to have the lowest losses due to the high collection network voltage, and the lowest cost due to the elimination of the collection platform (Lundberg, 2003). Several strings could be used in parallel to increase fault tolerance. The disadvantage of this system is that the transformer and converter in the turbine must be capable of isolating the full transmission voltage, and high voltage transformers with a low enough power rating are not commercially available. Several solutions to the converter and transformer problem have been proposed, for example by Parasai (2008).

Converters on platform

AC cluster

A further concept is the idea of connecting turbines with fixed-speed induction generators to a variable frequency AC collection grid, with strings of turbines being connected through a single converter. This places the converters on the collection platform, allowing them to be more easily repaired in the event of a fault, and a single large converter could potentially be cheaper than several small ones (Trilla, 2010). An AC or DC collection system could be used within the collection platform, as shown in Figure 5.21.

The speed of all the turbines in the string can be varied together to track the maximum power point for the current wind speed, but speed control over the individual turbines is lost. The speed of each turbine will be able to vary by a small amount relative to the others, due to the slip of the induction generator, with an increase in turbine speed leading to an increase in slip and an increase in

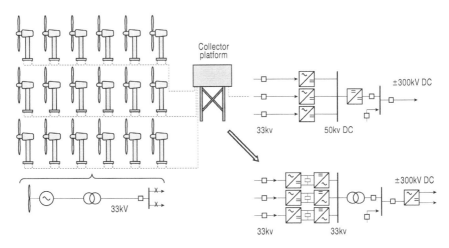

Figure 5.21 Cluster AC connection

torque. Depending on the number of turbines connected to each converter, this will result in a reduction in the amount of power extracted.

This system could also have an impact on the drive-train loads experienced by the turbines, as a turbine experiencing a gust would not be able to speed up to absorb the excess power, leading to a high transient torque, putting strain on the drive-train and blade roots. Research on the reliability of turbines in service has shown that the move to variable-speed turbines has reduced the level of blade failure compared with fixed-speed turbines.

Parallel DC cluster

This method, shown in Figure 5.22, uses a permanent magnet generator and passive rectifier in the turbine, with a DC-DC converter for each string of turbines. The speed of the turbine will be determined by the DC voltage of the string, so the system will behave in a similar way to the AC cluster connection system described previously, with similar issues of drive-train torque transients during gusts. It is considered that the passive rectifier will have considerably greater reliability than an active converter.

For a given DC voltage, the amount of possible speed variation of the turbine will depend on the generator inductance, with a higher inductance giving a greater variation in speed. The passive rectifier is unable to supply the generator with reactive power, and if the generator inductance is too high then the maximum torque will be reduced. Inductance is typically much higher in low speed machines, used in direct-drive turbines, and in these cases capacitors can be used between the generator and rectifier to supply the reactive power requirements.

The main advantage of DC over AC clustering is the greater efficiency of the permanent-magnet generator, compared with the induction generator used in the AC system. The greater current and voltage capability of the DC cables

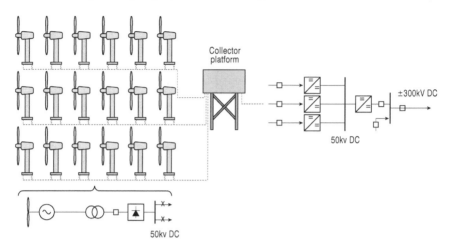

Figure 5.22 Parallel DC cluster connection

could also lead to larger cluster sizes, and a reduction in cable cost, but this could also reduce the power capture. A DC system could also reduce the number of conversion steps, increasing efficiency.

Series DC cluster

A variation of the parallel cluster arrangement is to connect the turbines in series, in a loop, with each loop controlled by a single converter, as shown in Figure 5.23. In this case, the converter will control the current within the loop, which will determine the generator torque within the turbine, and will be much more analogous to the conventional turbine control method. As the turbine speeds will be capable of varying individually, transient torque spikes should not be a problem, although this connection method has not been described in the literature, so the exact performance is unknown. Speed limitation for the turbines will need to be achieved using pitch control.

AC star

Star connection in this case involves connecting each turbine individually to a converter, and grouping converters together on a platform for easier access. It could also be possible to include redundant backup converters on the platforms for greater reliability. The converter platforms will output at an intermediate voltage, and an intermediate collection network will connect to the main collector platform for conversion to DC and transmission. This arrangement is shown in Figure 5.24.

For the basic AC star arrangement, conventional components are used. Permanent magnet generators can be produced which have an 11 kV output

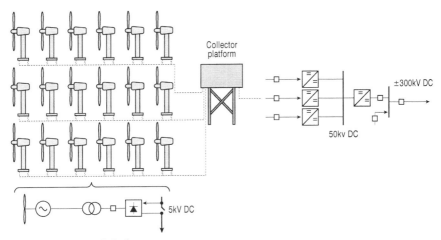

Figure 5.23 Series DC cluster connection

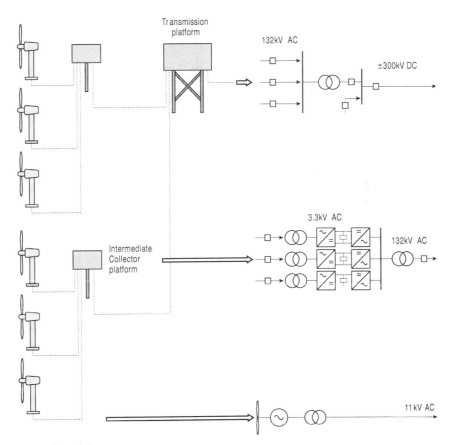

Figure 5.24 AC star connection

voltage, so an 11 kV collection network is used in order to eliminate the turbine transformers. 11 kV voltage source converters are not commonly available, so the collection voltage is stepped down to 3.3 kV for conversion.

This architecture could end up having a high cost, as it will require either very long cable lengths or a large number of intermediate platforms, both of which are costly.

DC star

Using DC within a star-type collection network could offer advantages in terms of cable cost and reduction in the number of conversion steps. This could involve connecting the turbines to the converter platform using AC, then having a DC connection to the transmission platform, as shown in Figure 5.25, or using passive rectifiers in the turbines and DC throughout. Fault protection in the higher voltage DC network could be difficult.

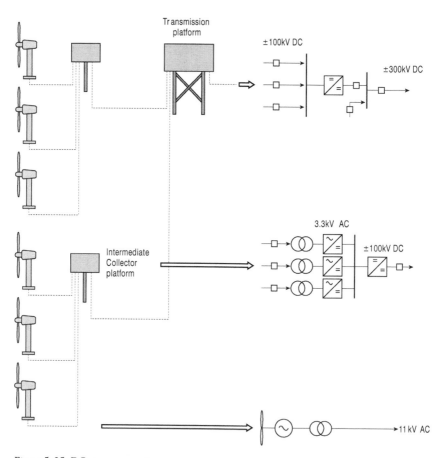

Figure 5.25 DC star connection

Star connection, no transmission platform

Similar to the DC system above, the transmission platform could be eliminated entirely if the collector platforms produce DC at the transmission voltage, as shown in Figure 5.26, although this would have all the problems associated with a high voltage DC network.

A multi-stage conversion process could be used to achieve the required voltage, or a series system could be used, in which the outputs of the individual turbine converters are connected in series. The latter option is similar to the series DC interconnection system, but will not suffer from the lack of tolerance for cable faults. However, it will still require converters featuring a high isolation voltage.

Cost

Calculations are based on a 1 GW wind farm, connected using a single voltage-source HVDC link at ±300 kV, this being the largest subsea cross-linked polyethylene (XLPE) cable capacity available at the time the study was performed. Larger wind farms, such as the UK Round 3 projects, could be built from multiple units of this size.

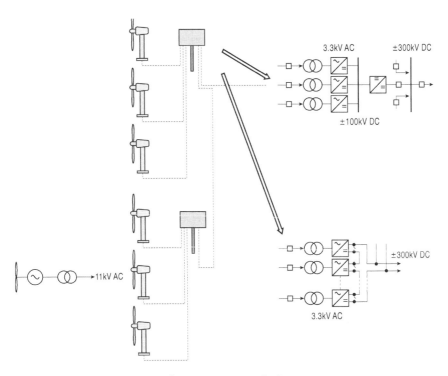

Figure 5.26 Star connection, without transmission platform

For the turbine, a rating of 10 MW is used, based on the trend towards larger turbines in offshore applications. While 10 MW turbines are not available at the time of writing, there are several designs in development. For the 10 MW turbine used, the blade length of the RePower 5M was scaled by √2 in order to double the swept area. To maintain the same tip speed ratio, the rated speed of the turbine was divided by the same factor. Two candidate scenarios from the converter on turbine and converter on platform cases were compared.

DC solutions have merit, though AC solutions are presently still favoured since they have greater operational history, and thus both operation and maintenance teams have more experience with AC solutions. However, looking at losses (Figure 5.27) DC would potentially have lower operational losses, and so this option may be seen in future as greater operational experience with DC is obtained.

Integration of energy storage

Electrical energy storage is an enabling technology permitting the management of intermittency, power quality and stability in large-scale power generation. Various electrical energy storage technologies are available, within the broad categories of chemical, mechanical and electrical technologies.

The supply of electrical power to a network of consumers involves continuously matching the supply of generated power to the load demand. This has to be achieved economically, while maintaining high levels of supply security and reliability, and high quality of supply in terms of voltage and frequency regulation. National power networks have generally been designed with large centralised generators, situated near fuel resource centres, with a high voltage transmission

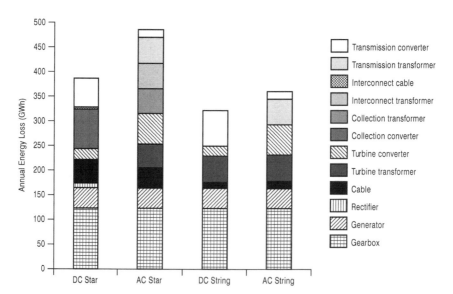

Figure 5.27 Breakdown of energy loss for the different configurations

grid and low voltage distribution (T&D) network to supply the loads. The overall demand is smoothed in the short term by diversity, with diurnal and seasonal variations, and is approximately predictable on an hourly basis. Demand prediction allows power generation plant to be optimally scheduled according to the plant economics and response time; and the demand deviations are balanced by governor control of on-line generators, together with fast-response generators and spinning reserve which can be brought on-line at short notice.

Although the energy source can be stored for some forms of renewable generation such as hydro and biomass, and power is despatchable according to load demand, other forms such as wind and solar photovoltaics are responsive to instantaneous meteorological conditions such as wind speed and solar irradiance. There is also likely to be increased generation by combined heat and power (CHP) systems, which offer efficiency improvements, improved economics, and reduction of CO_2 emissions, but are usually driven by heat demand, although this can be modified using a thermal store. In general it is desirable to be able to control generated power on demand, and the economics of electricity trading places a premium on predictability.

The nature of generation from dispersed renewable sources and CHP means that individual generators are relatively small (for example, less than 10 MW, although wind farms at the time of writing are increasing to 100s of MW) and connected into the distribution network, where the power is used locally. While penetration of distributed generation of the order of 10–20 per cent may be possible within existing T&D structures, it is widely believed that power systems will have to evolve to accommodate target levels distributed generation (Energy White Paper, DTI 2003). Existing supply networks are basically uni-directional, delivering power from a few central power generating nodes, through the T&D network branches to individual loads. A high level of distributed generation may require bi-directional power flow networks, with control of power routing. It is not yet clear how large existing networks will be adapted for new requirements, however this is being actively investigated, including the possible benefits of 'micro-grids' or 'quasi-autonomous operation', and on the other hand the possible opportunities for European-wide transmission and trading networks. Meanwhile, there is a pressing need for rural electrification schemes in developing countries, where multiple autonomous networks in geographically dispersed areas may be more economic than a single large interconnected grid network.

A simple calculation illustrates the challenge of incorporating high levels of renewable generation in the UK, where 10 per cent (or approx 40 TWh per annum) of electricity generation would mean 10–15 GW wind power capacity, assuming capacity factor in the range 30–45 per cent (Jenkins and Strbac, 2000). The minimum summer loading in the UK is less than 20 GW, and to ensure stability when maximum wind generation coincides with minimum loading it may at times be necessary to reduce wind generation output, or to trade on a large interconnected European grid, or to store bulk electrical energy. A full analysis of the impact and economics of storage in an electricity system requires knowledge of the storage technology performance and lifetime, which can be used in simulation models to consider various scenarios of generation mix and distribution network topology.

With the prospect of 40 GW offshore wind power capacity by 2050, it is increasingly clear that a range of balancing measures will be required, including demand-side management, electricity storage and interconnection of grid networks.

Possible scenarios for the configuration of offshore wind farms, with offshore wind farms and AC networks connected to shore using HVDC links (far offshore) and AC links (near offshore) could therefore benefit from the inclusion of storage.

Potential needs for storage include:

- To enhance stability in the offshore AC network in the steady state, and following external and internal disturbances and transients
- To avoid turbine trips due to offshore and onshore faults
- To avoid curtailment of wind farm output.

The storage requirements will include definition of:

- Power and energy
- Response time
- Charge and discharge rates
- Cycling lifetime
- Size, weight, and environmental specifications.

The requirements and applicable technologies can be summarised as in Table 5.1.

A hybrid technology solution for applications could be possible, to provide a range of cycling and power capability. A high energy solution for the curtailment application offers other services, e.g. arbitrage. A fast response solution for curtailment could also offer stability improvements. A priority is to maintain stability of the offshore wind farm network. Candidate technologies are in the 'power' storage category, i.e. supercapacitor, flywheels, batteries, SMES.

The study by Butler *et al.* (2002) suggests that management of renewable energy generation in electric power utility applications can potentially benefit from storage over a very wide spectrum of power (10 kW to 100 MW), storage

Table 5.1 Storage technology requirements by function

Application	Power	Storage time	Response	Cycling	Candidate technology	Location
Stability	Low–medium	Seconds	Sub-second	High	Super-capacitor Flywheel	Offshore
Turbine trips	High	Seconds–minutes	Sub-second	Low	Flywheel Battery	Offshore
Avoiding curtailment	High	Hours	Slow	Low	Battery Flow cell	Onshore

time (0.1 s to 10 h), and energy. A report by the European Commission (2001) suggests the integration of renewable energy into the electricity grid requires a range of storage systems with rated power > 10 kW, and storage times from around 0.1 hour up to 24 hours. The rated power and the storage time (discharge time at rated power) are key parameters, which categorise storage requirements and appropriate storage technologies. The ratio of rated power to energy capacity (P/E), or its inverse the storage time, leads to categories of 'high power' and 'high energy' storage, as summarised in Table 5.2.

Early studies by Davidson *et al.* (1980) reviewed the potential for large-scale bulk energy storage in the UK, while Infield (1984) studied the role of electricity storage and central electricity generation. The latter study concluded that there was an economic case for increased levels of efficient storage, and in addition transmission and distribution could be reduced if storage was distributed rather than centralised, although high storage efficiencies are required to realise significant savings.

At the 'power' end of the spectrum (storage times in the region of seconds to minutes), storage is applicable for power quality improvement, and for transmission grid stability applications. Power quality improvement measures applied at the distribution network level includes mitigation of voltage dips and sags caused by faults elsewhere in the transmission system, and reduction of voltage fluctuations (including 'flicker') and voltage rise caused by distributed generators. The wide spectrum of power and energy requirements means that a range of storage technologies may be needed.

Other key technical parameters include the cycling requirements (or lifetime energy throughput), the response time, and the in-out efficiency.

Electrical storage technologies

Electrical energy may be stored by transforming it into another form of energy, categorised into electrochemical, electromechanical, or electrical energy forms:

1   Electrochemical storage
    • Batteries (where the electrodes are part of the chemical reaction): lead-acid, lithium ion, nickel cadmium, sodium sulphur

Table 5.2 Storage duty – function and duty

Power / Energy [hour^{-1}]	Storage time	Applications
< 1 'Energy' storage	> 1 hour	• Load levelling • Peak shaving
60 to 1	1 minute to 1 hour	• Spinning reserve • Frequency and voltage regulation
> 60 'Power' storage	< 1 minute	• Power quality improvement • Transmission grid stability

- Flow cell (where the electrodes are catalysts for the chemical reaction): polysulphide bromide, vanadium, zinc bromine
- Hydrogen electrolyser and fuel cell

2   Electromechanical storage
- Pumped hydro
- Compressed air energy storage (CAES) and small compressed air storage (CAS)
- Flywheel

3   Electrical storage
- Superconducting magnetic energy storage (SMES)
- Supercapacitor (also considered as electrochemical storage).

The storage technologies have distinctly different characteristics, including the feasibility of implementation at various power or energy ratings, and they are at different states of technical and commercial maturity. Therefore it is possible to make only general observations of suitability for applications categorised according to storage times, as shown in Table 5.3. The information presented in Table 5.3 has been drawn from various sources, which also contain more details of the individual technologies (EC 2003, Butler *et al.* 2002, and Electricity Storage Association).

The overall cost of energy storage, including capital, operation and maintenance, and losses, is a key factor. The choice of technology also depends on the size of storage units, and the operational profile. Cost comparisons are difficult because costs may be quoted in terms of kW or kWh installed, while the operational cost per kWh throughput over the whole lifetime may be application dependent and is not quoted by manufacturers.

Batteries such as lead-acid and nickel cadmium are suitable for a wide range of applications, while advanced batteries with enhanced characteristics are under

Table 5.3 Storage physics and function

Technology:	Application:	'Energy' storage Load levelling Peak shaving	Spinning reserve Frequency and voltage regulation	'Power' storage Power quality improvement Transmission grid stability
	Storage time:	> 1 hour	1 minute to 1 hour	< 1 minute
Electro-chemical storage	Batteries	●	●	●
	Flow cells	●	○	
Electro-mechanical storage	Pumped hydro	●		
	CAES	●		
	Flywheel		○	●
Electrical storage	SMES		○	●
	Supercapacitor			●

development but have not reached the same level of maturity. Reviews of nine storage technologies, lead-acid battery; lithium battery; nickel batteries (nickel-zinc, nickel-cadmium, nickel metal-hydride); metal-air battery; electrolysers, hydrogen storage, and fuel cells; redox flow cells; flywheel; compressed-air; and supercapacitors, have been completed as part of the INVESTIRE network (EC 2003). The technology reports include a comparison of commercial systems and references for further information. The reports also include discussion of other factors that define performance, such as reliability, availability, ease of maintenance and service infrastructure, and safety.

Generic Wind Farm

As a basis for the research a generic scheme can be considered as shown in Figure 5.28. This includes an offshore AC network with one of more wind farms operating at 690 V and collection networks operating at 33 kV and 132 kV (for example), with transmission to shore via an HVDC VSC link operating at ±300 kV 1000 MW (for example).

Storage could be integrated at various points in the network, and the possible storage requirements and technologies can be summarised as in Table 5.4.

Dynamic simulation of electrical systems

While there are plenty of publications on basic MMC system modelling, system parameterisation continues to be less well discussed. This section covers the key

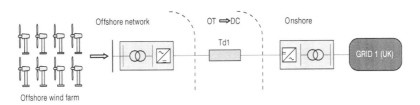

Offshore network   OT ⇒DC   Onshore

Td1

GRID 1 (UK)

Offshore wind farm

Figure 5.28 Example wind farm connection system

Table 5.4 Storage power and function

	Stability	Turbine trips	Avoiding curtailment
Power	Low-medium (per wind farm)	High (per turbine)	High
Storage time	Seconds	Seconds–minutes	Hours
Response	Sub-second	Sub-second	Minutes
Cycling	High	Low	Low
Storage technology	Supercapacitor, Flywheel, Battery		Flow cell, battery

elements and gives some general starting points. A typical radial link for the connection of a Round 3 offshore windfarm has a power rating of around 1000 MW at ± 300 kV according to National Grid's Offshore Development Information Statement (ODIS) (National Grid, 2011). Hence this MMC rating is our base-case.

MMC parameters

Number of MMC levels

A suitable starting point for designing/modelling an MMC is to determine the appropriate number of converter levels required. As the number of converter levels increases, the harmonic content of the output waveform decreases, but the computational efficiency of the MMC model also decreases when using traditional modelling techniques. Commercial MMCs contain hundreds of SMs per converter arm (Friedrich, 2010). The primary reason that such a large number of SMs per converter arm are required is to reduce the voltage stress across each SM to a few kV. It is however possible to use significantly less SMs and still not require AC filters. Determining the number of converter levels to model is therefore typically a compromise between the harmonic content of the output waveforms and the computational efficiency of the model.[1]

There are several harmonic limit standards in existence. These standards are defined by national and international bodies, as well as transmission and distribution network operators. There is however no set of common standards and the standards tend to be specific to a particular network and for a particular system voltage level. Harmonic limits may be defined for voltage waveform distortion, injected current, telephone-weighted voltage distortion and telephone-weighted current. Voltage waveform distortion limits are defined for the very high majority of networks, whereas the injected current limits are less widely used. Telephone-weighted limits are normally only considered when there is long exposure of telephone circuits to power circuits. The IEC 610003-3-6 and the IEEE 519 harmonic voltage limits are given in Table 5.5 and Table 5.6, respectively, from Alstom (2010).

Table 5.5 IEC 61000-3-6 harmonic voltage limits for high voltage systems

Odd harmonics non-multiple of 3		Odd harmonics multiple of 3		Even harmonics	
Harmonic order	Harmonic voltage (%)	Harmonic order	Harmonic voltage (%)	Harmonic order	Harmonic voltage (%)
5	2	3	2	2	1.4
7	2	9	1	4	0.8
11	1.5	15	0.3	6	0.4
13	1.5	21	0.2	8	0.4
17–49	–	21–45	0.2	10–50	–

Table 5.6 IEEE 519 harmonic voltage limits

Bus voltage	Individual voltage distortion	Total voltage distortion (THD) (%)
161 kV+	1	1.5

The IEC standard does not provide limitations for current injection or telephone-weighted values whereas the IEEE standard does.

For the test model used in this chapter a 31-level MMC was found to offer a good compromise by meeting the IEEE 519 harmonic voltage limits without having a significant impact on the simulation time.

Sub-module capacitance

The choice of the SM capacitance value is the next important parameter, and it is a trade-off between the SM capacitor ripple voltage and the size of the capacitor. A capacitance value which gives a SM voltage ripple in the range of ±5 per cent is considered to be a good compromise (Jacobson, 2010).

An analytical approach proposed by Marquardt *et al.* in Marquardt (2002) can be used to calculate the approximate SM capacitance required to give an acceptable ripple voltage for a given converter rating. This method effectively calculates the value of SM capacitance required for a given ripple voltage by determining the variation in the converter arm energy. This approach assumes that the output voltage and current is sinusoidal, that the DC voltage is smooth and split equally between the SMs, and that the converter is symmetrical. Circulating currents are also assumed to be zero. The key steps are presented here.

The current flowing through each converter arm contains a sinusoidal component, $I_a(t)/2$ and a DC component, I_g as shown in Figure 5.29 and described by:

$$I_{ua}(t) = I_g + \frac{1}{2}I_a(t) = I_g + \frac{1}{2}(\hat{I}_a \sin(\omega t + \varphi))$$

$$I_g = \frac{1}{3}I_{dc} \qquad m = \frac{\hat{I}_a}{2I_g}$$

$$\therefore I_{ua}(t) = \frac{1}{3}I_{dc}(1 + m\sin(\omega t + \varphi))$$

φ is the phase angle between $V_a(t)$ and $I_a(t)$. Neglecting the voltage drop across the arm impedance, each arm of the converter is represented as a controllable voltage source as described by:

$$V_{ua}(t) = \frac{1}{2}V_{dc} - \hat{V}_a(t) = \frac{1}{2}V_{dc} - \hat{V}_a \sin(\omega t) = \frac{1}{2}V_{dc}(1 - k\sin(\omega t))$$

$$k = \frac{2\hat{V}_a}{V_{dc}}$$

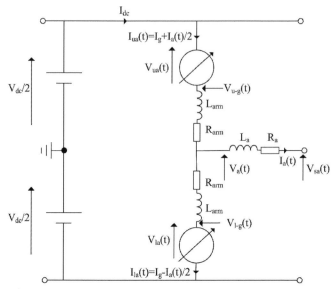

Figure 5.29 Single-phase equivalent circuit for MMC

The power flow of the upper phase arm, P_{ua}, is given by:

$$P_{ua}(t) = V_{ua}(t)I_{ua}(t) = \frac{1}{2}V_{dc}(1-k\sin(\omega t)) \times \frac{1}{3}I_{dc}(1+m\sin(\omega t + \varphi))$$

$$= \frac{V_{dc}I_{dc}}{6}(1-k\sin(\omega t))(1+m\sin(\omega t + \varphi))$$

where:

$$m = \frac{2}{k \times \cos\varphi}$$

The variation in stored energy for the upper arm over a half cycle, ΔW_{ua}, can be found by integrating P_{ua} between limits x_1 and x_2 (positive half cycle):

$$\Delta W_{ua} = \int_{x_1}^{x_2} P_{ua}(t) = \frac{P_d}{3\omega}m\left[1-\frac{1}{m^2}\right]^{3/2}$$

$$x_1 = -\varphi - \arcsin\left(\frac{1}{m}\right)$$

$$x_2 = \omega t = \pi + \arcsin\left(\frac{1}{m}\right) - \varphi$$

The required capacitance per SM for a ripple voltage factor ($0 > \varepsilon > 1$) can be determined as follows:

$$W_{Cap} = 0.5 \times C_{SM} \times V_{cap}^2 = \frac{\Delta W_{ua}}{4\varepsilon n} = \frac{\Delta W_{SM}}{4\varepsilon}$$

$$C_{SM} = \frac{\Delta W_{SM}}{2\varepsilon V_{Cap}^2}$$

This approach which was proposed by Marquardt *et al.* is widely used (Zhang, 2011; Ding, 2008). The SM capacitance calculation should only be used as an approximation as it is based on several assumptions (as described in the introduction of this section). The control strategy implemented for balancing the capacitor voltages will also have a significant impact on the capacitor ripple voltage. The SM's capacitance should be calculated based on the operating condition that causes the greatest variation in the capacitor's stored energy. This condition is likely to be when the converter is operating at maximum power.

Example SM capacitance calculation

The SM capacitance required to give a 10 per cent (±5 per cent) capacitor voltage ripple for a 1000 MW, ±300 kV, 31-level MMC is calculated below:

The average SM capacitor voltage can be calculated as follows:

$$V_{cap} = \frac{V_{dc}}{n} = \frac{600kV}{30} = 20kV$$

The DC current, AC phase voltage and current, and frequency are given by:

$$I_{dc} = \frac{P_d}{V_{dc}} = \frac{1000MW}{600KV} = 1670A$$

$$\hat{V}_a = 300kV$$

$$I_a = \frac{P_d/3}{V_a} = \frac{333MW}{212kV} = 1570A$$

$$\omega = 2\pi f = 314 rads$$

The values for I_g and m can be found from:

$$I_g = \frac{I_{dc}}{3} = \frac{1670A}{3} = 557A$$

$$m = \frac{\hat{I}_a}{2I_g} = \frac{\sqrt{2} \times 1570}{2 \times 557} = 2$$

The variation in the upper arm's stored energy is given by:

$$\Delta W_{ua} = \frac{1}{314} \frac{1000MW}{3} 2\left(1 - \frac{1}{2^2}\right)^{3/2} = 1.38MJ$$

Based on the assumption that the energy variation is shared equally between all SMs in the converter arm, the required SM capacitance for a voltage ripple of ±5 per cent is calculated by:

$$C_{SM} = \frac{1.38MJ / 30}{2 \times 0.05 \times 20kV^2} = 1150uF$$

According to Jacobson (2010) 30–40 kJ of stored energy per MVA of converter rating is sufficient to give a ripple voltage of 10 per cent (±5 per cent). Using this approximation, the value of SM capacitance can be calculated and compared with the above value. The value of SM capacitance based on 30 kJ/MVA is thus:

$$W_{SM} = \frac{30kJ \times 1000}{6n} = \frac{30MJ}{180} = 167kJ$$

$$C_{SM} = \frac{W_{SM}}{0.5 \times V_{Cap}^2} = \frac{167kJ}{0.5 \times 20kV^2} = 835\mu F$$

The value of SM capacitance based on 40 kJ/MVA is given by:

$$C_{SM} = \frac{222kJ}{0.5 \times 20kV^2} = 1110\mu F$$

The analysis conducted here shows that a value of 40 MJ of stored energy for a 1000 MVA converter gives a value of SM capacitance similar to the method proposed by Marquardt *et al.*.

Arm reactor

The arm reactors, also known as converter reactors, limb reactors and valve reactors, have two key functions. The first function is to limit the circulating currents between the legs of the converter, which exist because the DC voltages generated by each converter leg are not exactly equal. The second function is to reduce the effects of faults, both internal and external to the converter. By appropriately dimensioning the arm reactors, the circulating currents can therefore be reduced to low levels and the fault current rate of rise through the converter can be limited to an acceptable value.

The circulating current is a negative sequence (a-c-b) current at double the fundamental frequency, which distorts the arm currents and increases converter losses (Qingrui, 2010). For a typical MMC, the value of arm inductance required to limit the peak circulating current, $\hat{I}_{circ}$, can be calculated using the following equation from Qingrui (2010) as a starting point:

$$L_{arm} = \frac{1}{8\omega^2 C_{SM} V_{cap}} \left(\frac{S}{3\hat{I}_{circ}} + V_{dc}\right)$$

The second function of the arm reactor is to limit the fault current rate of rise to within acceptable levels. According to Dorn (2007), the Siemens HVDC Plus MMC convertor reactors limit the fault current to tens of amps per microsecond even for the most critical fault conditions, such as a short-circuit between the DC terminals of the converter. This allows the IGBTs in the MMC to be turned-off at non-critical current levels. Assuming that the DC voltage remains relatively constant from fault inception until the IGBTs in the converter are switched-off, the value of arm reactance required to limit the initial fault current rate of rise (dI_f/dt) can be described by:

$$L_{arm} = \frac{V_{dc}}{2(dI_f/dt)}$$

For example, the minimum value of arm reactance required to limit the fault current to 20 A/μs for a ±300 kV VSC during a terminal line-to-line DC fault is 15 mH. A 15 mH arm reactor would however result in very large circulating currents. Ideally the circulating current should be zero, however very large arm reactors, and/or SM capacitors, would be required for this to be achieved. The size of the arm reactor is therefore selected as a compromise between voltage drop across the arm reactor and the cost of the arm reactor, against the magnitude of circulating current. The circulating current can also be suppressed by converter control action or through filter circuits. For the test scenario used in this chapter a 45 mH (0.1 p.u.) arm reactor used in conjunction with the CCSC was found to offer a good level of performance.

Arm resistance

Accurate representation of converter losses for an MMC is beyond the scope of standard EMT type models. EMT type models often represent the conduction state of the SM's diodes/IGBTS using a fixed low-value resistance. The user can therefore select a resistance value to approximate converter losses. For example, the MMC model for the test system has an arm resistance of 0.9 Ω, which results in converter losses of approximately 0.5 per cent when operating at its rated power of 1000 MW.

Onshore AC network

For basic HVDC system studies the onshore network is often represented as an AC voltage source behind an impedance. The strength of an AC system is typically characterised by its short circuit ratio (SCR), where an SCR greater than 3 is defined as strong (Alstom, 2010).

$$SCR = \frac{V_n^2/Z_n}{P_{drated}}$$

For the test model used in this chapter, the 400 kV AC system was modelled to be relatively strong and highly inductive with an SCR of 3.5 and an X/R ratio of 20. The resistance and inductance values can therefore be calculated as follows:

$$Z_n = \frac{V_n^2}{SCR \times P_{drated}} = \frac{400kV^2}{3.5 \times 1000MVA} = 45.71\Omega$$

$$Z_n = \sqrt{R_n^2 + (20R_n)^2}$$

$$R_n = \sqrt{\frac{Z_n^2}{401}} = 2.28\Omega$$

$$L_n = \frac{20R_n}{\omega} = 0.145H$$

VSC-HVDC systems are presently symmetrical monopoles and typically employ a conventional AC transformer. The winding configuration of the converter transformer employed in this work is delta/star, with the delta winding on the converter side of the transformer as is the case for the Trans Bay Cable project (Friedrich, 2010). A tap-changer is used on the star winding of the transformers located onshore to assist with voltage regulation. The transformer leakage reactance is set to 0.15 p.u. with copper losses of 0.005 p.u. which are typical values for a power transformer (Grainger, 1994). A simplified diagram of the onshore system is shown in Figure 5.30.

DC system

Cable

The MMCs are connected by two 165 km HVDC cables with a nominal voltage and current rating of 300 kV and 1.7 kA respectively as per ODIS, which has now been replaced by National Grid's Electricity Ten Year Statement. Accurate models of the cables are required in order for the DC link dynamics of the scheme to be represented. The most common types of commercially available HVDC cable models are:

- PI-section model – This model lumps the cable's resistance, R, capacitance, C, and inductance, L, together and is normally adequate for steady-state simulations and for modelling short lengths of cable. As the frequency of interest

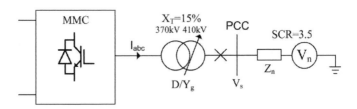

Figure 5.30 Onshore AC system

or the length of cable increases, more PI-sections are required to account for the distributed nature of the cable, which leads to additional computation time.

- Bergeron model – The Bergeron model is based on travelling wave theory and represents the distributed nature of the cable's LC parameters. The cable's resistance is lumped together and divided into three parts, 25 per cent at each end of the cable and 50 per cent in the middle. This model, similar to a PI-section, does not account for the frequency-dependent nature of the cable's parameters and is therefore essentially a single-frequency model.
- Frequency-dependent models – These models represent the cable as a distributed RLC model, which includes the frequency dependency of all parameters. This type of model requires the cable's material properties and geometry to be known. There are two frequency-dependent models available in, for example, PSCAD: the Frequency Dependent Mode Model (FDMM) and the Frequency Dependent Phase Model (FDPM). The key difference between the two models is that the mode model does not represent the frequency-dependent nature of the internal transformation matrix, whereas the phase model does through direct formulation in the phase domain.

In Beddard (2014) a coupled equivalent PI model, Bergeron model, FDMM and FDPM were compared in terms of accuracy and simulation speed for a range of studies. The results show that the type of cable model can have a significant impact on the model's overall response for typical VSC-HVDC studies. The results also show that the efficiency of the Bergeron, FDMM and FDPM were similar while the use of a coupled equivalent PI model resulted in a longer simulation time. It is for these reasons that the FDPM, which is the most accurate model, is used for the example test model.

DC braking resistor

DC braking resistors are normally required on VSC-HVDCs schemes used for the connection of windfarms (Jiang-Häfner and Ottersten, 2009). There are situations, such as an onshore AC grid fault, which diminish the onshore converter's ability to export the energy from the windfarm. The bulk of this excess energy is stored in the scheme's SM capacitors leading to a rise in the DC link voltage. The DC braking resistor's function is to dissipate the excess energy and to therefore prevent unacceptable DC link voltages.

The worst-case scenario is where the onshore MMC is unable to effectively export any active power. This can occur for severe AC faults such as a solid three-phase to ground fault at the PCC, as shown in Figure 5.31. The braking resistor should therefore be rated to dissipate power equal to the windfarm power rating.

A DC braking resistor can be modelled simply as a voltage-dependent current source (Mohammadi, 2013), or as a power electronic switch connected in series with resistor for a more detailed representation. The basic control strategy for a DC braking resistor is to turn the IGBT valve on once the DC voltage has exceeded a particular threshold and to turn it off once the DC voltage has

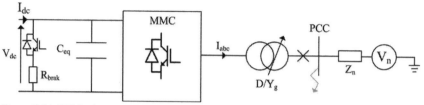

Figure 5.31 DC braking resistor

returned below a specific threshold (Jiang-Hafner and Ottersten, 2009), however more complex strategies can be employed (Livermore, 2013). A DC braking resistor could also be used to assist the surge arrestors in limiting transient DC over-voltages. The DC braking resistors in the test system used in this chapter employ the basic upper and lower threshold control and are designed to limit the maximum DC voltage to 1.2 p.u. using a 500 Ω resistor. The IGBT braking valve would therefore be required to conduct up to 1500 A.

Example: Point-to-point VSC-HVDC link for Round 3 windfarm

The modelling outlined in the previous sections of the chapter was used to produce an MMC VSC-HVDC link for a typical Round 3 offshore windfarm as shown in Figure 5.32. No storage is incorporated.

Example results are given to give the reader a basis for comparison with model results they produce when implementing the methods introduced in this chapter.

Start-up

The onshore converter's SM capacitors are energised from the onshore AC system as shown in Figure 5.33. The converter is initially in the blocked state, which effectively forms a 6-pulse bridge enabling each arm of the converter to charge-up. During this initial charging phase, resistors are inserted between the AC system and the converter to prevent excessive in-rush current. In this model the rectified DC voltage is not equal to the nominal DC voltage (600 kV), and therefore at 0.3 s, the converter is operated in DC voltage control to obtain the nominal DC voltage. During the entire charging process the offshore converter's circuit breakers are open and the SM capacitors are charged from the DC voltage created by the onshore converter. Once the charging process is complete the offshore AC circuit breakers are closed.

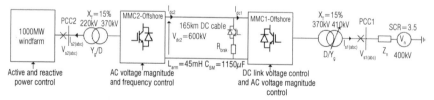

Figure 5.32 MMC VSC-HVDC link for Round 3 windfarm

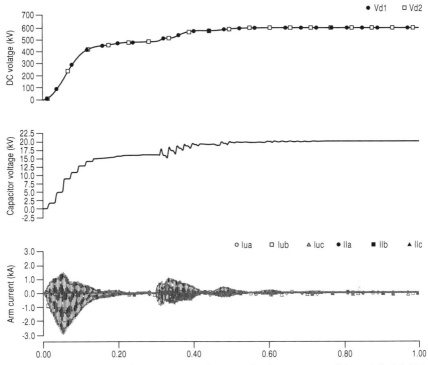

Figure 5.33 Start-up procedure; capacitor voltages are for the upper arm of phase A for MMC1

Windfarm power variations

At 1 s, the windfarm is ordered to inject 1 GW of active power, the order is then reduced to 500 MW at approximately 3.1 s and again increased to 750 MW at approximately 4 s. The onshore converter is initially set to supply 330 MVAr to the onshore grid and is then set to absorb 330 MVAr at approximately 2.1 s while operating at maximum active power. The converter is therefore able to meet the required reactive power demands set out in the GB grid code (leading and lagging power factor of 0.95) (National Grid, 2013). Figure 5.34 clearly shows that the VSC-HVDC link is capable of responding to the power demands of the windfarm.

The link's steady-state response for the windfarm operating at maximum power is shown in Figure 5.35. The phase voltages and phase currents at the PCC for the onshore network ($V_{s1(abc)}$, $I_{s1(abc)}$) and the offshore network ($V_{s2(abc)}$, $I_{s2(abc)}$) are shown to be highly sinusoidal and as such have a small harmonic content. The DC voltages are smooth, while the DC current exhibits a small ripple of approximately ±1.5 per cent of the nominal value. The DC current ripple can be reduced further by disabling the CCSC, however, this would increase converter losses.

The total harmonic distortion of the line-to-line voltages at the onshore PCC is shown in Figure 5.36. This analysis confirms that the harmonic content is within the 1.5 per cent limit set out in the IEEE 519 standards.

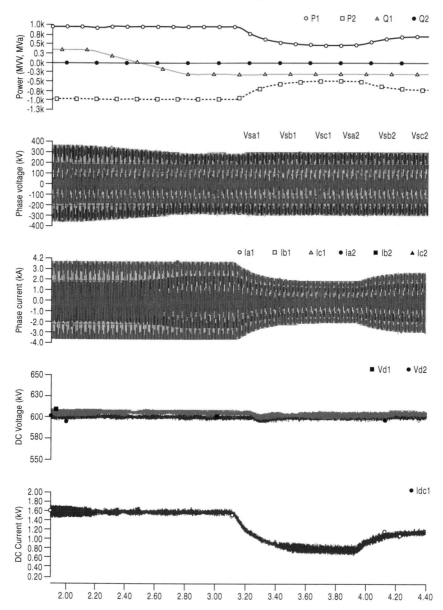

Figure 5.34 Link response to variations in windfarm power; *x*-axis – time (s)

MMC *detailed performance*

The output phase voltages ($V_{1(abc)}$), arm currents ($I_{u(abc)}$, $I_{l(abc)}$) and difference currents ($I_{diff(abc)}$) for the onshore converter operating at 1000 MW are shown in Figure 5.37. The staircase voltage waveform produced by the MMC is more

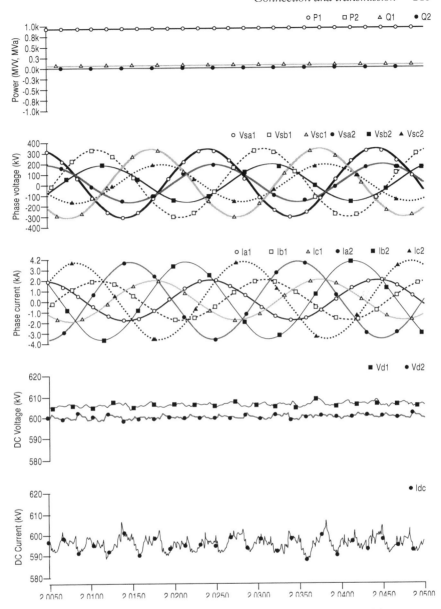

Figure 5.35 Link response at steady-state for Pw = 1000 MW; *x*-axis – time (s)

evident than the voltages measured at the PCC. The CCSC is able to suppress the circulating current to very small values, as is evident in the bottom graph of Figure 5.37, since there is virtually no AC component. The absence of the circulating current component in the arm current ensures that there is little distortion and that losses are minimised. The effect of the CCSC on the waveforms can be seen by comparing Figure 5.37 with Figure 5.38.

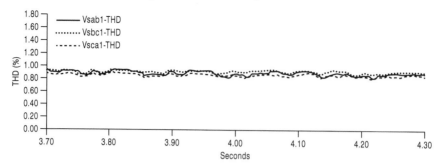

Figure 5.36 THD for the line-to-line voltages at PCC1 for Pw = 1000 MW; x-axis – time (s)

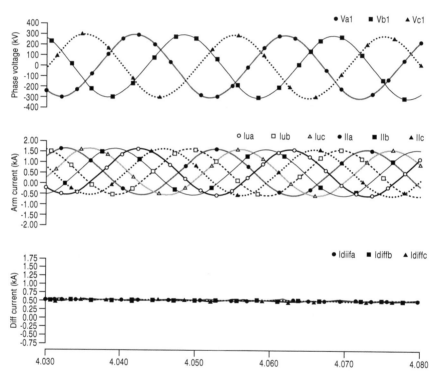

Figure 5.37 Phase voltages, arm currents and difference currents for onshore MMC with Pw = 1000 MW; x-axis – time (s)

The effect of the CCSC on the converter losses can be assessed by measuring the rms value of the arm current. Comparing Figure 5.39 with Figure 5.40 shows that disabling the CCSC increases the arm current by approximately 25 per cent, indicating a significant increase in the converter losses. This increase in arm current may also require the valve components to be designed for a higher current rating.

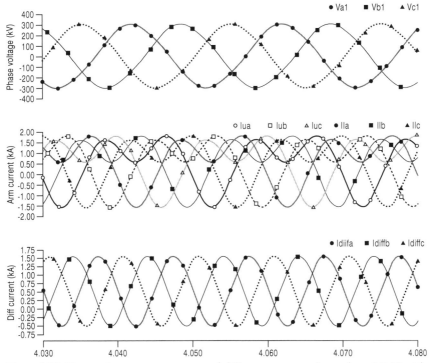

Figure 5.38 Phase voltages, arm currents and difference currents for onshore MMC with Pw = 1000 MW and CCSC disabled; *x*-axis – time (s)

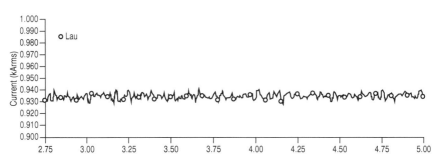

Figure 5.39 RMS value of the upper arm current for phase A with Pw = 1000 MW and CCSC enabled; *x*-axis – time (s)

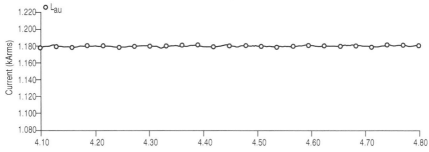

Figure 5.40 RMS value of the upper arm current for phase A with Pw = 1000 MW and CCSC disabled; *x*-axis – time (s)

Note

1 The number of converter levels modelled does not have a significant impact on the model's efficiency when using some types of MMC model, such as the detailed equivalent model (Gnanarathna, 2011).

References

Alstom, 2010 "HVDC – Connecting to the future", Alstom.

Antonopoulos, Angquist, L. and Nee, N.P., 2009, "On dynamics and voltage control of the Modular Multilevel Converter," in 13th European Conference on Power Electronics and Applications, 2009. EPE '09., pp. 1–10.

Barnes, M. and Beddard, A., 2012, "Voltage Source Converter HVDC Links – The state of the Art and Issues Going Forward," *Energy Procedia*, pp. 108–122.

Beddard, A. and Barnes, M., 2014, "HVDC cable modelling for VSC-HVDC applications," PES General Meeting 2014, IEEE, 2014, pp. 1–5.

Butler, P., Miller, J.L., and Taylor, P.A., 2002, *Energy Storage Opportunities Analysis Phase II Final Report, A study for the DOE energy storage systems program*, Sandia National Laboratories, report SAND2002-1314.

Crine, J.P., 1998, *Electrical, Chemical and Mechanical Processes in Water Treeing*, IEEE Transactions on Dielectrics and Electrical Insulation, Vol. 5.

Davidson B.J., Glendenning I., Harman R.D., Hart A.B., Maddock B.J., Moffit R.D., Newman V.G., Smith T.F., Worthington P.J., and Wright J.K., 1980, "Large-scale electrical energy storage", *IEEE Proceedings Part A*, vol. 127, no. 6, July. pp. 345–385.

Ding, G., Ding, M. and Tang, G., 2008, "An Innovative Hybrid PWM Technology for VSC in Application of VSC-HVDC Transmission System," IEEE Electrical Power & Energy Conference.

Dorn, J., Huang, H., and Retzmann, D., 2007, "Novel Voltage-Sourced Converters for HVDC and FACTS Applications," Cigre.

DTI, 2003, "Our Energy Future: Creating a Low-Carbon Economy", Energy White Paper no. 68, Cm5761 London: Department of Trade and Industry.

EC, 2003. "Storage Technology reports", EC INVESTIRE network project ENK5-CT-2000-20336, http://www.itpower.co.uk/investire

Electricity Storage Association, "Storage Technology Comparison", http://www.electricitystorage.org/technology/storage_technologies/technology_comparison

Friedrich, K., 2010, "Modern HVDC PLUS application of VSC in Modular Multilevel Converter topology," 2010 IEEE International Symposium on Industrial Electronics (ISIE), pp. 3807–3810.

Gnanarathna, U.N. et al., 2010, *Modular Multi-Level Converter Based HVDC System for Grid Connection of Offshore Wind Power Plant*, IET ACDC, London, pp. 1–5.

Gnanarathna U.N. et al., 2011, "Efficient Modeling of Modular Multilevel HVDC Converters (MMC) on Electromagnetic Transient Simulation Programs," *IEEE Transactions on Power Delivery*, vol. 26, pp. 316–324.

Grainger, J. and Stevenson, W., 1994, *Power Systems Analysis* McGraw-Hill.

Infield, D.G., 1984, "A Study of Electricity Storage and Central Electricity Generation", Rutherford Appleton Laboratory Technical Report RAL-84-045, 1984. http://epubs.stfc.ac.uk/work-details?w=50349

Jacobson, B., Karlsson, P., Asplund, G., Harnefors, L. and Jonsson, T., 2010, "VSC-HVDC Transmission with Cascaded Two-Level Converters," Cigre B4-110.

Jenkins, N., and Strbac, G., 2000, "Increasing the value of renewable sources with energy storage", in *Renewable Energy Storage*, IMechE Seminar Publication No. 2000-7, London: Institution of Mechanical Engineers, 1–10.

Jiang-Häfner, Y and Ottersten, R, 2009, "HVDC with Voltage Source Converters – A Desirable Solution for Connecting Renewable Energies," 8th International Workshop on Large-Scale Integration of Wind Power into Power Systems and Transmission Networks for Offshore Wind Farms. Bremen, Germany, October 2009

Jun, L. et al., 2009, "Control of multi-terminal VSC-HVDC transmission for offshore wind power," 13th European Conference on Power Electronics and Applications, 2009. EPE '09., pp. 1–10.

Lesnicar, A. and Marquardt, R., 2003, "An innovative modular multilevel converter topology suitable for a wide power range," in Power Tech Conference Proceedings, 2003 IEEE Bologna, Vol.3, p. 6.

Livermore, L., 2013, "Integration of Offshore Wind Farms Through High Voltage Direct Current Networks," PhD Thesis, Cardiff University.

Lundberg, S., 2003, "Performance comparison of wind park configurations", Chalmers University of Technology.

Marquardt, R., Lesnicar, A. and Hildinger, J., 2002, "Modulares Stromrichterkonzept für Netzkupplungsanwendung bei hohen Spannungen bei hohen Spannungen," *ETG-Fachbericht*, vol. 88, pp. 155–161.

Merlin, M.M.C et al., 2014, "The Alternate Arm Converter: A New Hybrid Multilevel Converter With DC-Fault Blocking Capability," *IEEE Transactions on Power Delivery*, vol. 29, pp. 310–317.

Meyer, C., Kowal, M. and De Doncker, R.W., 2005, "Circuit breaker concepts for future high-power DC applications", 40th IAS Industry Applications Conference.

Mohammadi, M. et al., 2013, "A study on Fault Ride-Through of VSC-connected offshore wind farms," Power and Energy Society General Meeting (PES), 2013 IEEE, pp. 1–5.

Monjean, P. et al., 2010, "Innovative DC connections for offshore wind and tidal farms", EPE Wind Energy Chapter Symposium, Stafford, UK.

National Grid, 2011, "Offshore Development Information Statement" Company Report.

National Grid, 2013, "The Grid Code – Issue 5 Rev 3".

Parasai, A. et al., 2008, "A New Architecture for Offshore Wind Farms", *IEEE Transactions on Power Electronics*, vol. 23, pp. 1198–1204.

Qingrui T. and Zheng, X., 2011, "Impact of Sampling Frequency on Harmonic Distortion for Modular Multilevel Converter," *IEEE Transactions on Power Delivery*, vol. 26, pp. 298–306.

Qingrui, T., Zheng, X. and Jing, Z., 2010 "Circulating current suppressing controller in modular multilevel converter," IECON 2010 – 36th Annual Conference on IEEE Industrial Electronics Society, pp. 3198–3202.

Saeedifard M. and Iravani, R., 2010, "Dynamic Performance of a Modular Multilevel Back-to-Back HVDC System," *IEEE Transactions on Power Delivery*, vol. 25, pp. 2903–2912.

Sannio, A., Breder, H. and Koldby Nielsen, E., 2006, "Reliability of Collection Grids for Large Offshore Wind Parks", 9th International Conference on Probabilistic Methods Applied to Power Systems, Stockholm, Sweden.

Trilla, L. et al. , 2010, "Analysis of total power extracted with common converter wind farm topology", 19th International Conference on Electrical Machines, ICEM.

Yang, J., Fletcher, J. and O'Reilly, J., 2010, "Multiterminal DC Wind Farm Collection Grid Internal Fault Analysis and Protection Design", *IEEE Transactions on Power Delivery*, vol. 25, pp. 2308–2318.

Zhan, C. et al., 2010, "DC Transmission and Distribution System for a Large Offshore Wind Farm", 9th IET International Conference on AC and DC Power Transmission.

Zhang, Y. et al., 2011, "Analysis and Experiment Validation of a Threelevel Modular Multilevel Converters," International Conference on Power Electronics.

Zhong, D., Tolbert, L.M., and Chiasson, J.N., 2006, "Active harmonic elimination for multilevel converters," *IEEE Transactions on Power Electronics*, vol. 21, pp. 459–469.

6   Wind turbine control

William E. Leithead and Adam Stock

Wind turbines have control systems whose primary purpose is to keep the turbine operating states, i.e. the rotor speed, aerodynamic torque, generated power etc., at their wind speed dependent design values. However, as the size of wind turbines has increased, the demands on the controller have increased and its design task has become more demanding. For modern multi-MW turbines, the controller objectives now commonly include the reduction of structural loads, specifically, the loads on the tower and blades.

Wind energy installed capacity in the UK is becoming significant and is expected to continue to grow, in part through offshore development of large-scale wind farms. One of the consequences of the resulting high levels of grid-penetration would be that these offshore farms will need to operate in a manner similar to conventional generation; that is, they will need to become integrated power plants. To do so, individual turbines will need to operate much more flexibly and not simply output the level of power as dictated by the wind. A wind farm level controller is required to make the farm mimic conventional generation, or at least, some aspects of conventional generation and enable flexible operation of individual turbines.

In SUPERGEN Wind, both the design of advanced wind turbine controllers with the objective of reducing the tower and blade loads and the design of wind farm controllers have been explored. Both are considered in the following. The performance of these controllers is assessed using simulation models of three-bladed variable speed wind turbines, the SUPERGEN 5MW Exemplar Wind Turbine, the SUPERGEN 2MW Exemplar Wind Turbine and a 1.5MW wind turbine based on the latter. These are referred to here as the 5M, 2MW and 1.5MW wind turbine, respectively.

Turbine control schemes for reduced dynamic loading

In above rated wind speeds, the basic objective of the full envelope controller for a variable speed wind turbine is to maintain the rotor torque and speed at their rated values, i.e. to reject the disturbances on rotor torque and speed due to the constantly varying wind speed. To achieve the former, the power converter is exploited to provide a fixed generator reaction torque. Since the power converter is very fast acting, open loop control suffices. To achieve the latter, the pitch angles of the wind turbine blades are adjusted in response to a measurement of rotor or generator speed. All three blades are adjusted in unison; that is, assuming the rotor has three blades, the angles of all three blades are kept the same. Accordingly, in above rated wind speed, the wind turbine full envelope controller typically has a structure as depicted in Figure 6.1, where ω_{g0} and T_{g0} are the generator speed and torque set points, ω_g and T_g are the actual generator speed and torque, β_d is the demanded blade pitch angle and T_d is the demanded generator torque. As the power converter is relatively fast acting, it can be assumed that the fluctuations in T_g relative to T_{g0} are very small and that the response of T_g to perturbations in β_d is very weak. Indeed, when $T_d = T_{g0}$, T_g can essentially be assumed to be equal to T_d and so to T_{g0}, as in Figure 6.1. However, the speed control loop is much slower acting and the fluctuations in ω_g relative to ω_{g0} are relatively much greater. The magnitude of the variations in generator speed is, of course, dependent on the bandwidth of the speed control loop, which is typically 1 rad/s.

In this section, the term *basic controller* is used when referring to the speed controller with the structure in Figure 6.1. When the controller has the additional objectives of reducing tower and/or blade loads, the basic controller is augmented by additional feedback loops acting on pitch demand.

Tower load reduction

The usual approach to tower load reduction is to augment the pitch demand by an additive adjustment in response to a measurement of the tower head fore-and-aft velocity or acceleration (Bossanyi 2003a, Leithead 2005). In effect there are two pitch feedback loops, an inner tower feedback loop (TFL) and an outer rotor speed feedback loop. Neglecting the open loop generator torque control, the structure of this augmented controller with both a speed feedback loop and a tower feedback loop is depicted in Figure 6.2, where $\dot{\varphi}_T$ is the tower head fore-and-aft velocity. Essentially this additional feedback loop acts to increase tower motion damping. When applied to the 2MW wind turbine, the TFL controller in comparison to the basic controller alone achieves a reduction of 6.67 per cent in the lifetime Damage Equivalent Loads (DELs).

An alternative approach to reducing the tower loads (Leithead 2006, Chatzopoulos 2010a, 2010b) is based on the observation that increased tower loads

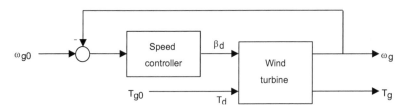

Figure 6.1 Above rated wind turbine controller

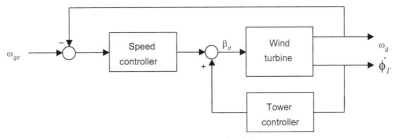

Figure 6.2 Augmented controller with tower feedback loop

are caused by the increased pitch activity that accompanies tighter regulation of generator speed. Consequently, since the major contribution to the tower loads is at frequencies close to the first fore-and-aft tower frequency, typically around 2 rad/s, reducing pitch angle adjustments due to the generator speed control action, in the vicinity of the tower frequency would reduce these loads. Using a simple notch filter alone would compromise the speed control performance. Instead, the speed control is enacted through a combination of pitch and torque demand as in Figure 6.3. The element Y is chosen to reduce pitch activity in the vicinity of the tower frequency; for example, it could be chosen to be a low pass filter or a notch filter. The element X is chosen such that the transmittance from its input to ω_g is similar to the transmittance from β_d to ω_g. Hence, the controller C in Figure 6.3 can remain unchanged from the controller used with the basic controller, Figure 6.1. Here, the acronym CCD is used when referring to the speed controller with the structure in Figure 6.3.

For any particular wind speed time series, the generator speed obtained using the CCD controller is the same as that obtained using the basic controller. However, because the gain from T_d to ω_g is much weaker than from β_g to ω_g, there can be large generator torque and power fluctuations relative to their rated values. These fluctuations directly impact on the drive-train components such as the gearbox and generator and should be avoided.

To reduce the impact on drive-train components, the structure of the above rated wind turbine controller is first modified such that the speed control loop is replaced by a power control loop as in Figure 6.4 (Chatzopoulos 2010b, 2011), where P_0 is the generated power set point and P is the actual generated power.

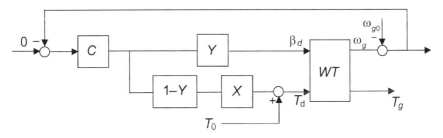

Figure 6.3 Modified speed control loop

Since the fluctuations in T_g about T_{g0} are relatively small in comparison to the fluctuations in ω_g about ω_{g0}, and indeed can be considered essentially zero, power control, Figure 6.4, is exactly equivalent to speed control, Figure 6.1.

Since the response of T_g to perturbations in β_d is very weak, the fluctuations in P relative to P_0 arise essentially from fluctuations in ω_g. Hence, if P is well controlled then so is ω_g, $(\omega_g - \omega_0)(T_g - T_{g0})$ is negligible and the plant output, $P = \omega_g T_g$, can be replaced by $\omega_{g0} T_{g0} + (\omega_g - \omega_{g0})T_0 + (T_g - T_{g0})\omega_{g0}$, see Figure 6.5(a). Equivalently, P can be replaced by $\omega_{g0} + (\omega_g - \omega_{g0}) + (T_g - T_{g0})\omega_{g0}/T_{g0}$, see Figure 6.5(b). Furthermore, the speed controller in Figure 6.5(b) can remain unchanged from the controller in Figure 6.1.

The modification to the speed control loop depicted in Figure 6.3 for tower load reduction, can also be applied to the power based controller in Figure 6.5(b); that is, the control strategy in Figure 6.5(b) is modified to that of Figure 6.6(a). As before, the element Y is chosen to reduce pitch activity in the vicinity of the tower frequency. The element X is chosen such that the transmittance from its input to the output of the feedback loop, i.e. $(\omega_g - \omega_{g0}) + (T_g - T_{g0})\omega_{g0}/T_{g0}$, is similar to the transmittance from β_d to $(\omega_g - \omega_{g0}) + (T_g - T_{g0})\omega_{g0}/T_{g0}$. Hence, the controller C in Figure 6.6(a) can still remain unchanged from that used in Figure 6.1.

For any particular wind speed time series, the generated power obtained using the control strategy in Figure 6.6(a) is the same as that obtained using the control

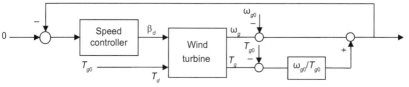

Figure 6.4 Above rated power based controller

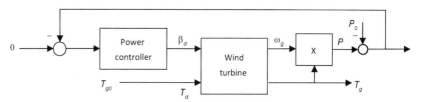

a) Controller with P replaced by $\omega_{g0} T_{g0} + (\omega_g - \omega_{g0}) T_{g0} + (T_g - T_{g0})\omega_{g0}$

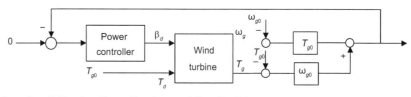

b) Controller with P replaced by $\omega_{g0} + (\omega_g - \omega_{g0}) + (T_g - T_{g0})\omega_{g0}/T_{g0}$

Figure 6.5 Equivalent power based controllers

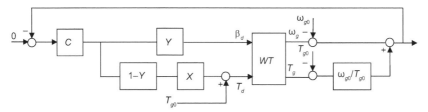

a) Modified power-based controller

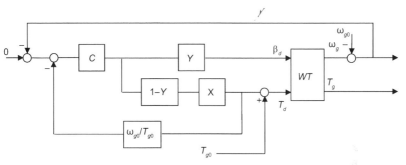

b) Equivalent modified speed controller

Figure 6.6 Power based controller for load reduction

strategy in Figure 6.4 and, hence, in Figure 6.1; that is, it is power that continues to be regulated.

The response of T_g to perturbations in ω_g and ω_g to perturbations in T_d are both weak, for the latter in the sense that the system acts like a low pass filter due to the wind turbine rotor inertia being very large. Hence, while regulating well the output of the power control loop in Figure 6.6(a), it is not possible to have large perturbations in T_g, albeit accompanied by large counteracting fluctuations in ω_g, except at frequencies significantly lower than the tower frequency. However, at these lower frequencies, the power control loop output is essentially regulated through pitch action alone as the element Y is chosen to be either a low pass filter or a notch filter. Consequently, since the closed loop bandwidth for the power control loop is 1 rad/s, the fluctuations in T_g arising from the power control loop are small. Of course, in terms of speed regulation, the control structure in Figure 6.6(a) is slightly less effective than that of Figure 6.3 and that of Figure 6.1 as these fluctuations are accompanied by small counteracting fluctuations in ω_g.

The power based controller structure in Figure 6.6(a) can be reformulated into an equivalent speed based controller structure, see Figure 6.6(b). Here, the acronym PCC is used when referring to the controller with the structure in Figure 6.6(a) or its equivalent Figure 6.6(b).

Other than a strongly varying aerodynamic gain that can be counteracted very effectively by global gain-scheduling (Leith 1996), the wind turbine dynamics are essentially linear. Hence, for the controllers in Figures 6.3 and 6.6, all the control elements, C, X and Y, can be presumed linear. The application of these controllers to the 2MW wind turbine is discussed below. All the results presented are obtained from a Bladed simulation of this turbine.

The element Y for the controllers, in both Figures 6.3 and 6.6, is chosen to be a notch filter centred on 2.47 rad/s, the tower fore-and-aft frequency for the 2MW wind turbine; that is:

$$Y = \frac{s^2 + 1.4s + 6.1}{s^2 + 3.2s + 6.1} \tag{6.1}$$

For the CCD controller, Figure 6.3, the element X is chosen to be:

$$X = \frac{1.68 \times 10^7}{s^3 + 19.17s^2 + 122.5s^1 + 260.9} \tag{6.2}$$

The Bode plots for the open loop systems in Figure 6.1, the basic controller strategy, and Figure 6.3, the CCD controller strategy, at 16 m/s wind speed are compared in Figure 6.7(a). Other than in the vicinity of the tower frequency, the two transmittances are very similar as required. Nevertheless, the impact of the tower can be clearly observed. In the Bode plot for the basic controller open loop system, there is a rapid 360° phase loss near the tower frequency due to the presence of a pair of lightly damped right half-plane zeros. The element X in Figure 6.3, the CCD controller, has enabled these zeros to be replaced by a pair of left half-plane zeros, thereby improving the stability margins. For the PCC controller, Figure 6.6, the element X is chosen to be:

$$X = \frac{-8.73 \times 10^6}{s^4 + 30.1s^3 + 303s^2 + 1030s + 100} \tag{6.3}$$

The open loop systems in Figure 6.1, the basic controller, and Figure 6.6, the PCC controller, are compared in Figure 6.7(b). The same observations as made previously with regard to Figure 6.7(a), apply.

The CCD control scheme is applied to the 2MW wind turbine. In comparison to the basic controller it achieves a reduction of 7.13 per cent in the DELs of the tower, a little better than the TFL applied to the basic controller as in Figure 6.2. As expected, the generator speed with the CCD controller is indistinguishable from that of the basic control scheme, see Figure 6.8(a), in which the generator speeds obtained with both are compared at a mean wind speed 16 m/s. However, as discussed, the CCD controller induces large drive-train torque excursions, see Figure 6.8(b). Applying the TFL in combination with the CCD controller achieves a reduction of 10.86 per cent in the DELs of the tower, considerably better than the TFL alone.

The PCC control scheme is also applied to the 2MW wind turbine. In combination with a TFL, it achieves a reduction of 10.11 per cent in the DELs of the tower in comparison to the basic controller, again considerably better than the TFL alone. As expected, the generator power with the PCC controller is indistinguishable from that of the basic control scheme, see Figure 6.9(a) in which the generator powers obtained with both are compared at a mean wind speed of 14 m/s. Furthermore, as discussed, the PCC controller does not cause any

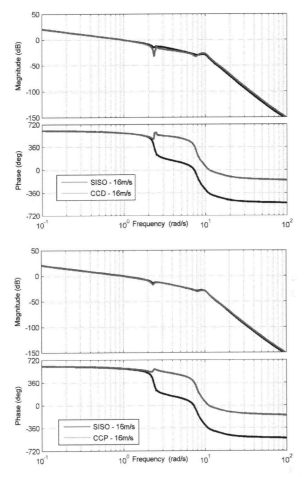

Figure 6.7 Open loop Bode plots for open loop systems in Figures 6.1, 6.3 and 6.6 at 16 m/s

significant increase in drive-train torque excursions in comparison to the basic controller, see Figure 6.9(b).

The performance of all controllers, basic, CCD and PCC, are summarised in Table 6.1. Both the CCD and PCC controllers are very effective at regulating the

Table 6.1 Tower LEDs with basic, CCD and PCC control schemes

Control scheme	Lifetime equivalent tower loads	Load reduction (%)
Basic	7.2186e+06	–
Basic + TFL	6.7369e+06	6.67
CCD	6.7036e+06	7.13
CCD + TFL	6.4341e+06	10.86
PCC + TFL	6.4891e+06	10.11

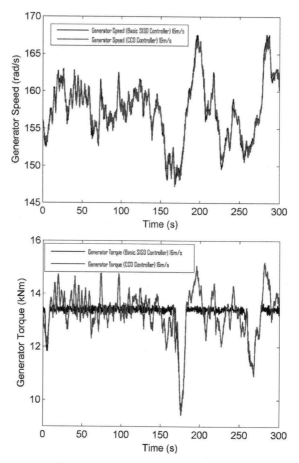

Figure 6.8 Comparison of basic and CCD controllers

DELs of the tower, achieving a significantly better reduction than that achieved by the tower feedback loop alone. Further benefits are increased stability margins, due to the removal of right half-plane zeros, and a reduction in pitch activity. Unlike the CCD controller, the PCC controller does not cause any marked increase in drive-train loads and so is preferred.

Blade load reduction

In large multi-MW wind turbines, each blade has its own actuator capable of adjusting the blade to some demanded pitch angle. The basic speed controller in Figure 6.1 provides the same demanded pitch angle to each actuator, thus, causing all three blades to pitch in unison. However to reduce the blade and hub loads, each turbine blade must be pitched independently; that is, each actuator must be provided with a different demanded pitch angle, usually through additive

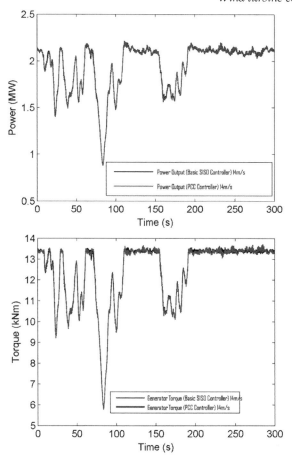

Figure 6.9 Comparison of basic and PCC controllers (14 m/s)

adjustments to the demanded pitch angle from the basic speed controller. In this context, the basic speed controller is here referred to as the central controller.

Neglecting the open loop generator torque control, the usual approach to reducing the blade and hub loads by independently pitching the blades (Bossanyi 2003b, 2005) is to augment the speed control loop in Figure 6.1 with additional feedbacks acting in response to measurements of the out-of-plane root bending moments, M_1, M_2 and M_3, for blades 1, 2 and 3, respectively, as shown in Figure 6.10(a). The out-of-plane blade root bending moment is the preferred measurement to other root bending moments since it is minimally impacted on by gravity. The structure of the central controller in Figure 6.10(a) is depicted in Figure 6.10(b). The bending moments, M_1, M_2 and M_3, are transformed to M_d and M_q using the Coleman transform

$$\begin{bmatrix} M_d \\ M_q \end{bmatrix} = \frac{2}{3} \begin{bmatrix} \cos(\theta_R) & \cos(\theta_R + 2\pi/3) & \cos(\theta_R + 4\pi/3) \\ \sin(\theta_R) & \sin(\theta_R + 2\pi/3) & \sin(\theta_R + 4\pi/3) \end{bmatrix} \begin{bmatrix} M_1 \\ M_2 \\ M_3 \end{bmatrix} \quad (6.4)$$

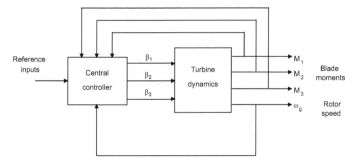

a) Augmented controller with blade bending moments feedbacks

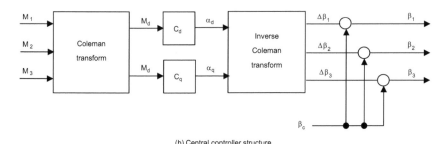

(b) Central controller structure

Figure 6.10 Independent pitch control

where θ_R is the azimuth angle of the rotor. Essentially M_d and M_q can be interpreted as the overturning or nodding moment and the yawing moment on the rotor. Two quasi-pitch angles, α_d and α_q, are demanded by the controllers, C_d and C_q, and are transformed to $\Delta\beta_1$, $\Delta\beta_2$ and $\Delta\beta_3$ using the inverse Coleman transform

$$\begin{bmatrix} \Delta\beta_1 \\ \Delta\beta_2 \\ \Delta\beta_3 \end{bmatrix} = \begin{bmatrix} \cos(\theta_R) & \sin(\theta_R) \\ \cos(\theta_R + 2\pi/3) & \sin(\theta_R + 2\pi/3) \\ \cos(\theta_R + 4\pi/3) & \sin(\theta_R + 4\pi/3) \end{bmatrix} \begin{bmatrix} \alpha_d \\ \alpha_q \end{bmatrix} \tag{6.5}$$

The $\Delta\beta_1$, $\Delta\beta_2$ and $\Delta\beta_3$ are additive adjustments to β_c, which is the collective pitch demand from the speed controller. Typically, C_d and C_q are chosen to be the same and to be PI (proportional-integral) controllers with gains that are scheduled on pitch angle. The strategy depicted in Figure 6.10 is referred to here as independent pitch control (IPC).

An alternative approach to reducing the blade and hub loads is to enclose each blade within its own feedback (Neilson 2010, Leithead 2009a, 2009b, 2009c) that adjusts the pitch angle of that blade in response to a measurement of its root bending moment as shown in Figure 6.11. These three additional moment feedback loops are independent of each other and of the basic speed control loop. The strategy depicted in Figure 6.11 is referred to here as independent blade control (IBC).

A simplified representation of the basic speed control loop, Figure 6.1, is depicted in Figure 6.12(a) with, for clarity, only one of the three actuators

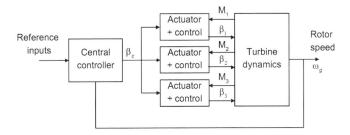

Figure 6.11 Independent blade control

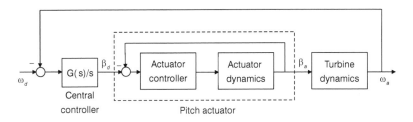

a) Speed control loop and pitch actuator

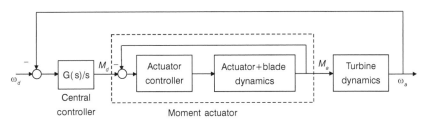

b) Speed control loop and moment actuator

Figure 6.12 Speed control loop actuators

explicitly shown. The pitch actuator itself has a controller, represented in Figure 6.12(a) by a feedback loop, which causes the actual pitch angle β_a to follow a demanded pitch angle β_d. Since, as shown explicitly in Figure 6.12(a), the central controller includes integral action, the actual generator speed ω_a is driven to follow the demanded generator speed ω_d by means of the pitch actuator. In IBC, the monotonic relationship between pitch angle and root bending moment is exploited to introduce a moment feedback enclosing the actuator as in Figure 6.12(b). In effect, this inner feedback loop converts the pitch actuator into a moment actuator that causes the actual blade root bending moment M_a to follow a bending moment demanded by the central controller M_d. Since the central controller continues to include integral action, the actual generator speed ω_a is still driven to follow the demanded generator speed ω_d.

The speed control loop in Figure 6.12(a) is subject to a generator speed disturbance induced by wind speed fluctuations. Within its closed loop bandwidth, the feedback acts to reduce these disturbances. Similarly, both the speed control loop and the moment control loop in Figure 6.12(b) are subject to disturbances induced by wind speed fluctuations, see Figure 6.13; that is, the outer feedback loop continues to be subject to a generator speed disturbance whilst the inner feedback loop is now subject to blade root bending moment disturbance. Potentially, both disturbances on the generator speed and on the bending moment can be reduced, the former by the speed control loop, the latter by the moment control loop. For the inner feedback loop, the disturbances manifest themselves as peaks in the bending moment spectral density function at frequencies of 1Ω and 2Ω, where Ω is the angular velocity of the rotor, typically about 2 rad/s for large wind turbines. These spectral peaks are caused by the blade rotating through the wind field at the rotor. Consequently, other than a phase shifts of 120° and 240°, there is a fair degree of correlation between the disturbances on each blade. Hence, the combined action of all three moment controllers can, by design, be made substantially independent of the speed control loop, particularly when the former is focused on the 1Ω and 2Ω peaks. For the outer feedback loop, the disturbances are those due to wind turbulence over frequencies up to 1 rad/s. However, the design of the blade root bending moment controller is not independent of the design of the generator speed controller, since they are coupled through the turbine structural dynamics. Furthermore, the dynamics on which the design of the moment controller depends, are a priori highly nonlinear and dependent on the azimuthal angle of the rotor.

Consider the dynamics of the blade defined with respect to axes with origin at the centre of rotation of the rotor and aligned with the hub; that is, the root of the blade is stationary with respect to these blade axes. However, both the blade and the axes move linearly with the tower head, move rotationally with the nacelle and rotate with the rotor. Hence, these axes define a non-inertial reference frame for analysis of the blade and actuator dynamics. If the blade root were not attached to the wind turbine but instead was stationary, say, with the axes fixed relative to earth axes, then these axes would define an inertial reference frame. The dynamics in the two cases are very different but there is a simple way to relate them. The Newtonian dynamics of any system with respect to a non-inertial reference

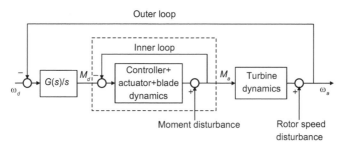

Figure 6.13 Disturbances acting on the moment feedback loop

frame are the Newtonian dynamics with respect to an inertial reference frame with the addition of fictitious forces (Feynman 2006) proportional to the relative acceleration of the reference frames. A well-known example of a fictitious force is centrifugal force. This relationship between the dynamics in non-inertial and inertial reference frames is illustrated by the simple system in Figure 6.14.

The system in Figure 6.14 consists of a mass, m, and spring with stiffness, k. The spring is attached at one end to a mounting fixed with respect to the xy axes set in Figures 6.14(a) and (b). In Figure 6.14(a), the xy axes are stationary and so define a reference frame with respect to which Newton's laws can be directly applied to derive the dynamics of the mass. The equation of motion for the mass, m, is

$$m\ddot{x} = -k(x - x_0) \tag{6.6}$$

where x is the displacement of the mass and x_0 is the displacement of the mass when in equilibrium. In Figure 6.14(b), the xy axes set is non-stationary and so does not define a reference frame with respect to which Newton's laws can be applied. Instead, the dynamics of the mass must be derived with respect to the reference frame defined by the stationary axes set, $x_i y_i$. The displacement of the xy axes set with respect to the $x_i y_i$ axes set is x_R and the displacement of the mass with respect to the $x_i y_i$ set is x_i and x_{i0} when in equilibrium. The equation of motion for the mass, m, is now

$$m\ddot{x}_i = -k(x_i - x_{i0}) \tag{6.7}$$

$$\Rightarrow m\ddot{x}_i = m(\ddot{x} + \ddot{x}_R) = -k(x_i - x_{i0}) = -k((x + x_R) - (x_0 + x_R))$$

$$\Rightarrow m\ddot{x} = -k(x - x_0) + F_f \tag{6.8}$$

where

$$F_f = -m\ddot{x}_R \tag{6.9}$$

Hence, (6.7), the equation of motion with respect to the reference frame defined by the stationary axes set $x_i y_i$, is the same as (6.8); that is, (6.6) the equation of motion with respect to the reference frame defined by the stationary

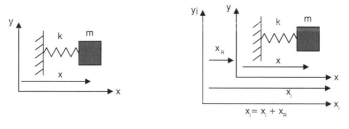

a) Inertial reference frame   b) Non-inertial reference frame

Figure 6.14 Dynamics of simple system in inertial and non-inertial reference frames

axes set *xy* in Figure 6.14(a) but with the addition of the fictitious force F_f. From (6.9), it can be seen that the fictitious force is simply the product of the mass and the acceleration of the stationary axes set xy_i relative to the non-stationary axes set *xy* in Figure 6.14(b). In Figure 6.14(b) and in similar contexts, since the *xy* axes define a non-stationary reference frame, the relationship (6.8) is referred to as the dynamics with respect to the non-inertial reference frame whilst, since it corresponds to the situation in Figure 6.14(a) with the *xy* axes defining a stationary reference frame, the relationship (6.8) with $F_f = 0$, i.e. the relationship (6.6), is referred to as the dynamics with respect to the inertial reference frame.

The above reference frame based transformation of the Newtonian dynamics can be applied to the moment feedback loop of Figure 6.13. With respect to a non-inertial reference frame fixed with respect to the hub, the plant dynamics consist of the actuator dynamics, blade dynamics and the dynamics of the rest of the wind turbine to which it couples. With respect to the inertial reference frame, the plant dynamics consist of the actuator dynamics and blade dynamics only, see Figure 6.15(a).

Consider a single non-rigid blade with fixed pitch angle on a wind turbine with non-rotating rotor and with rigid tower and nacelle when the reference frame fixed relative to the hub is an inertial one. A simple Newtonian dynamics lumped parameter model of the blade (Leithead 2009a) is

$$\begin{bmatrix} \ddot{\theta}_B \\ \ddot{\phi}_B \end{bmatrix} = -\frac{1}{J}\begin{bmatrix} M_{IP} \\ M_{OP} \end{bmatrix} + \frac{1}{J}\begin{bmatrix} M_{A\theta_B} \\ M_{A\phi_B} \end{bmatrix} \tag{6.10}$$

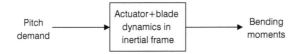

a) Plant dynamics defined with respect to inertial reference frame

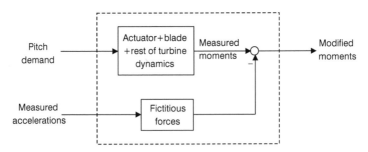

b) Plant dynamics defined with respect to non-inertial reference frame

Figure 6.15 Equivalent plant dynamics for moment feedback loop

where J is inertia of the blade relative to the blade root, θ_B and φ_B are, respectively, the blade in-plane and out-of-plane angular displacements, M_{IP} and M_{OP} are, respectively, the in-plane and out-of-plane root bending moments and $M_{A\theta_B}$ and $M_{A\varphi_B}$ are, respectively, external in-plane and out-of-plane torques relative to the blade root. The in-plane and out-of-plane root bending moments are given by

$$
\begin{bmatrix} M_{IP} \\ M_{OP} \end{bmatrix} = J \begin{bmatrix} \omega_E^2 \cos^2(\beta) + \omega_F^2 \sin^2(\beta) & -(\omega_E^2 - \omega_F^2)\sin(\beta)\cos(\beta) \\ -(\omega_E^2 - \omega_F^2)\sin(\beta)\cos(\beta) & \omega_E^2 \sin^2(\beta) + \omega_F^2 \cos^2(\beta) \end{bmatrix} \begin{bmatrix} \theta_B \\ \varphi_B \end{bmatrix} \quad (6.11)
$$

where β is the blade pitch angle and ω_E and ω_F are, respectively, the blade edge and flap frequencies.

Now consider the same blade but with the rotor rotating and with non-rigid tower and nacelle when the reference frame is no longer an inertial one. The difference in the Newtonian dynamics from (6.10) (Leithead 2009a, 2009c) can be represented by fictitious forces, more correctly in this case, by fictitious torques; that is,

$$
\begin{bmatrix} \ddot{\theta}_B \\ \ddot{\varphi}_B \end{bmatrix} = -\frac{1}{J}\begin{bmatrix} M_{IP} \\ M_{OP} \end{bmatrix} + \frac{1}{J}\begin{bmatrix} M_{A\theta_B} \\ M_{A\varphi_B} \end{bmatrix} + \frac{1}{J}\begin{bmatrix} M_{F\theta_B} \\ M_{F\varphi_B} \end{bmatrix} \quad (6.12)
$$

where $M_{F\theta_B}$ and $M_{F\varphi_B}$ are, respectively, fictitious in-plane and out-of-plane torques relative to the blade root. In (6.12), the external in-plane and out-of-plane torques are the aerodynamic moments.

It follows from above that, by subtracting fictitious forces or, more correctly in this case, fictitious torques from the measured blade root bending moment as in Figure 6.15(b), the plant dynamics of the moment feedback loop defined with respect to the non-inertial reference frame are transformed to those defined with respect to the inertial reference frame, Figure 6.15(a); that is,

$$
\begin{bmatrix} M_{IP} \\ M_{OP} \end{bmatrix}\Bigg|_{\text{Measured}} \Rightarrow \begin{bmatrix} M_{IP} \\ M_{OP} \end{bmatrix}\Bigg|_{\text{Measured}} - \begin{bmatrix} M_{F\theta_B} \\ M_{F\varphi_B} \end{bmatrix} \quad (6.13)
$$

In this way, the dynamics of the moment control loop and the design of its controller have been totally isolated from the rest of the turbine dynamics and the turbine central controller. Because this transformation is based solely on the relationship between the dynamics in non-inertial and inertial reference frames, it is valid for all representations of the dynamics and not solely for that of (6.10) and (6.11). Hence, this decoupling is very robust.

The motion of the blade does not have 6 degrees of freedom but is constrained through it being attached to the hub and the nacelle being attached to the tower. In determining the fictitious torques for a blade, cognisance must be taken of this constraint. Furthermore, the nacelle and hub are subject to both rotational and linear motion. Hence, the blade fictitious torques must include a contribution due to the former, the *angular fictitious torques*, and a contribution due to the latter, the *linear fictitious torques*. To discuss the derivation of blade fictitious torques several axes sets are required, specifically:

- *Earth axes* (X_E-Y_E-Z_E): the origin coincides with the base of the tower, the X_E-axis points vertically upwards, the Y_E-axis and Z_E-axis are horizontal with the Z_E-axis pointing upwind with the axis of rotation of the hub lying in the $X_E Z_E$ plane.
- *Tower axes* (X_T-Y_T-Z_T): the origin coincides with the centre of rotation of the rotor, the axes are parallel to the earth axes.
- *Rotor axes* (X_R-Y_R-Z_R): the origin coincides with the centre of rotation of the rotor, the Z_R-axis is the axis of rotation of the rotor, the X_R and Y_R axes rotate with the drive shaft, the X_R-axis is normal to the blade root cross-section.
- *Blade body centred axes* (X_B-Y_B-Z_B): the origin coincides with the origin of the rotor axes and the axes are aligned with principal axes of the blade such that they coincide with the rotor axes when the blade is straight.

Note that, when defining the tower axes and rotor axes, the tower is assumed sufficiently axially stiff that twisting of the tower need not be considered and, when defining the blade body centred axes set, the blade is additionally assumed sufficiently axially stiff that twisting of the blade also need not be considered. Furthermore, the pitch angle is fixed. The notations used for the different axes sets are defined in Table 6.2. For rotation R_2 the yaw and pitch Euler angles are θ and φ, and for rotation R_2' the yaw and pitch Euler angles are θ' and φ'. Since both the tower and blade are axially stiff, no Euler roll angles are required. Suffices x, y and z, are used to indicate the individual components of vectors.

For angular fictitious torques, the non-inertial reference frame for the blade is defined by the moving rotor axes whilst the inertial reference frame is defined by

Table 6.2 Notations for the axes sets for angular fictitious torques

$\tilde{\Omega}$	Angular velocity components of the body centred axes relative to the earth axes in body centred axes coordinates
$\tilde{\Omega}_R$	Angular velocity components of the rotor axes relative to the earth axes in rotor axes coordinates
$\tilde{\Omega}'$	Angular velocity components of the rotor axes relative to the earth axes in body centred axes coordinates
$\tilde{\omega}_B$	Angular velocity components of the body centred axes relative to the rotor axes in body centred axes coordinates
r_B	Displacement relative to origin of the tower axes of the blade centre of mass in tower axes coordinates
$\ddot{r}_{To}$	Linear acceleration relative to the earth axes of the tower axes in tower axes coordinates
R_2	Euler angle (3–2–1) rotation from the rotor axes to the blade body centred axes
R_2'	Euler angle (3–2–1) rotation from the tower axes to the blade body centred axes

the rotor axes but with them stationary, i.e. with the tower and nacelle stationary and the hub not rotating. With respect to body centred axes coordinates, the rotational Newtonian dynamics of the blade in the inertial reference frame are

$$I_B \dot{\tilde{\omega}}_B + \tilde{\omega}_B \times (I_B \tilde{\omega}_B) = \tilde{M}_E \tag{6.14}$$

where I_B is the inertia matrix of the blade and $\tilde{M}_E$ is the external torque applied to the blade in body centred axes coordinates. The rotational Newtonian dynamics in the non-inertial frame are

$$I_B \dot{\tilde{\Omega}} + \tilde{\Omega} \times (I_B \tilde{\Omega}) = \tilde{M}_E \tag{6.15}$$

Substituting $\tilde{\Omega} = \tilde{\Omega}' + \tilde{\omega}_B$ in (6.15)

$$I_B \dot{\tilde{\Omega}} + \tilde{\Omega} \times (I_B \tilde{\Omega}) = I_B (\dot{\tilde{\Omega}}' + \dot{\tilde{\omega}}_B) + (\tilde{\Omega}' + \tilde{\omega}_B) \times (I_B (\tilde{\Omega}' + \tilde{\omega}_B)) = \tilde{M}_E \tag{6.16}$$

Hence,

$$I_B \dot{\tilde{\omega}}_B + \tilde{\omega}_B \times (I_B \dot{\tilde{\omega}}_B) = \tilde{M}_E - I_B \dot{\tilde{\Omega}}' - \tilde{\Omega}' \times (I_B \tilde{\Omega}') - \tilde{\Omega}' \times (I_B \tilde{\omega}_B) - \tilde{\omega}_B \times (I_B \tilde{\Omega}') \tag{6.17}$$

and the angular fictitious torques applied to the blade are

$$\tilde{M}_{FR} = -I_B \dot{\tilde{\Omega}}' - \tilde{\Omega}' \times (I_B \tilde{\Omega}') - \tilde{\Omega}' \times (I_B \tilde{\omega}_B) - \tilde{\omega}_B \times (I_B \tilde{\Omega}') \tag{6.18}$$

Since

$$I_B \approx \begin{bmatrix} 0 & 0 & 0 \\ 0 & J & 0 \\ 0 & 0 & J \end{bmatrix} \tag{6.19}$$

the angular fictitious torques become

$$\tilde{M}_{FR} = -I_B \dot{\tilde{\Omega}}' - J \begin{bmatrix} 0 \\ -\tilde{\Omega}'_x \tilde{\Omega}'_z + \tilde{\omega}_{Bx} \tilde{\Omega}'_z + \tilde{\omega}_{Bz} \tilde{\Omega}'_x \\ -\tilde{\Omega}'_x \tilde{\Omega}'_y + \tilde{\omega}_{Bx} \tilde{\Omega}'_y + \tilde{\omega}_{By} \tilde{\Omega}'_x \end{bmatrix} \tag{6.20}$$

The two terms $J\tilde{\Omega}'_x \tilde{\Omega}'_z$ and $J\tilde{\Omega}'_x \tilde{\Omega}'_y$ can be interpreted as centrifugal stiffening of the blade and no further consideration of them is required. Since the magnitudes of $\tilde{\omega}_{Bx}$, $\tilde{\omega}_{By}$, $\tilde{\omega}_{Bz}$, $\tilde{\Omega}'_x$ and $\tilde{\Omega}'_y$ are all small and products of these components can be neglected,

$$\tilde{\omega}_B = \begin{bmatrix} -\dot{\theta}\sin(\varphi) \\ \dot{\varphi} \\ \dot{\theta}\cos(\varphi) \end{bmatrix} \text{ and } I_B \dot{\tilde{R}}_2 \tilde{\Omega}_R = I_B \dot{\tilde{R}}_2 R_2^{-1} \tilde{\Omega}' \approx -J \begin{bmatrix} 0 \\ \tilde{\omega}_{Bx} \tilde{\Omega}'_z \\ 0 \end{bmatrix} \tag{6.21}$$

where

$$\tilde{\Omega}' = R_2\tilde{\Omega}_R \; ; \; R_2 = \begin{bmatrix} \cos(\theta)\cos(\varphi) & \sin(\theta)\cos(\varphi) & -\sin(\varphi) \\ -\sin(\theta) & \cos(\theta) & 0 \\ \cos(\theta)\sin(\varphi) & \sin(\theta)\sin(\varphi) & \cos(\varphi) \end{bmatrix} \quad (6.22)$$

It follows that in blade body centred axes coordinates

$$\tilde{M}_{FR} \approx -I_B R_2 \dot{\tilde{\Omega}}_R - I_B \dot{R}_2 \tilde{\Omega}_R - J \begin{bmatrix} 0 \\ \tilde{\omega}_{Bx}\tilde{\Omega}'_z \\ 0 \end{bmatrix} \approx -I_B R_2 \dot{\tilde{\Omega}}_R = -R_2(I_R \dot{\Omega}_R) \quad (6.23)$$

where I_R is the blade inertia matrix in rotor axes coordinates; that is,

$$I_R = R_2^{-1} I_B R_2 \quad (6.24)$$

Since both θ and φ are small, $I_R \approx I_B$. Hence, the angular fictitious torques in the rotor frame are $-I_B \dot{\Omega}_R$.

For linear fictitious torques, the non-inertial reference frame for the blade is defined by the moving tower axes whilst the inertial reference frame is defined by the tower axes but with them stationary, i.e. with the tower and nacelle stationary. Since the location of the blade centre of mass in body centred coordinates is $[l \quad 0 \quad 0]^T$,

$$r_B = R_2'^{-1}\begin{bmatrix} l \\ 0 \\ 0 \end{bmatrix} = \begin{bmatrix} \cos(\theta')\cos(\varphi')l \\ \sin(\theta')\cos(\varphi')l \\ -\sin(\varphi')l \end{bmatrix} \; ;$$

$$R_2'(m_B r_B \times \ddot{r}_B) = m_B l^2 \begin{bmatrix} 0 \\ \ddot{\varphi}' + \dot{\theta}'^2\sin(\varphi')\cos(\varphi') \\ \ddot{\theta}'\cos(\varphi') - 2\dot{\varphi}'\dot{\theta}'\sin(\varphi') \end{bmatrix} \quad (6.25)$$

where

$$R_2' = \begin{bmatrix} \cos(\theta')\cos(\varphi') & \sin(\theta')\cos(\varphi') & -\sin(\varphi') \\ -\sin(\theta') & \cos(\theta') & 0 \\ \cos(\theta')\sin(\varphi') & \sin(\theta')\sin(\varphi') & \cos(\varphi') \end{bmatrix} \quad (6.26)$$

Furthermore, the linear kinetic energy of the blade is

$$T = \frac{1}{2}m_B \dot{r}_B^2 = \frac{1}{2}m_B l^2(\dot{\varphi}'^2 + \dot{\theta}'^2\cos^2(\varphi')) \quad (6.27)$$

and

$$\frac{d}{dt}\left(\frac{\partial T}{\partial \dot{\theta}'}\right) - \frac{\partial T}{\partial \theta'} = m_B l^2(\ddot{\theta}'\cos(\varphi') - 2\dot{\theta}'\dot{\varphi}'\sin(\varphi'))\cos(\varphi')$$

$$\frac{d}{dt}\left(\frac{\partial T}{\partial \dot{\varphi}'}\right) - \frac{\partial T}{\partial \varphi'} = m_B l^2(\ddot{\varphi}' + \dot{\theta}'^2\sin(\varphi')\cos(\varphi'))$$

$$(6.28)$$

From Lagrangian dynamics considerations, $m_B l^2 (\ddot{\varphi}' + \dot{\theta}'^2 \sin(\varphi') \cos(\varphi'))$ and $m_B l^2 (\ddot{\theta}' \cos(\varphi') - 2\dot{\theta}' \dot{\varphi}' \sin(\varphi')) \cos(\varphi')$ are the rates of change in angular momentum of the blade about the Y_B-axis and Z_B-axis, respectively. The components in $R_2' (m_B r_B \times \ddot{r}_B)$ only differ from those of (6.28) by a $\cos(\varphi')$ term, which is essentially 1 since φ' is a small angle. It follows that the components in $m_B r_B \times \ddot{r}_B = \dfrac{d}{dt} (m_B r_B \times \dot{r}_B)$ are themselves the rates of change in angular momentum of the blade about the X_B-axis, Y_B-axis and Z_B-axis, respectively, transformed into tower axes coordinates. Consistent with blade moment of inertia about the X_B-axis being negligible, see (6.19), and the tower and blade being sufficiently axially stiff that their twisting need not be considered, the first component in $R_2' (m_B r_B \times \ddot{r}_B)$, i.e. the torque about the X_B-axis, is zero.

With respect to tower axes coordinates, the linear Newtonian dynamics of the blade in the inertial reference frame are

$$m_B \ddot{r}_B = F_E - 2\lambda r_B \tag{6.29}$$

where m_B is the blade mass and F_E is the external force applied to the blade in tower axes coordinates. The Lagrange multiplier term, $2\lambda r_B$ is included to meet the constraint that $(r_B^T \cdot r_B)$ is constant. Equivalent to (6.29), rotational Newtonian dynamics with respect to the centre of rotation of the blade are

$$m_B r_B \times \ddot{r}_B = r_B \times F_E - 2\lambda r_B \times r_B = r_B \times F_E \tag{6.30}$$

The linear Newtonian dynamics in the non-inertial frame are

$$m_B (\ddot{r}_{To} + \ddot{r}_B) = F_E - 2\lambda r_B \tag{6.31}$$

and the rotational Newtonian dynamics equivalent to them are

$$m_B r_B \times \ddot{r}_B = r_B \times F_E - 2\lambda r_B \times r_B - m_B r_B \times \ddot{r}_{To} = r_B \times F_E - m_B r_B \times \ddot{r}_{To} \tag{6.32}$$

From the observation that the components in $m_B r_B \times \ddot{r}_B$ are rates of change in the angular momentum of the blade about the X_B-axis, Y_B-axis and Z_B-axis, respectively, transformed into tower axes coordinates and from comparison of (6.32) to (6.30), the corresponding fictitious torques in tower axes coordinates are

$$
\begin{aligned}
M_{FL} &= -m_B r_B \times \ddot{r}_{To} = -m_B \begin{bmatrix} \cos(\theta')\cos(\varphi')l \\ \sin(\theta')\cos(\varphi')l \\ -\sin(\varphi')l \end{bmatrix} \times \begin{bmatrix} \ddot{x}_{To} \\ \ddot{y}_{To} \\ \ddot{z}_{To} \end{bmatrix} \\
&= -m_B l \begin{bmatrix} \sin(\varphi')\ddot{y}_{To} + \sin(\theta')\cos(\varphi')\ddot{z}_{To} \\ -\sin(\varphi')\ddot{x}_{To} - \cos(\theta')\cos(\varphi')\ddot{z}_{To} \\ -\sin(\theta')\cos(\varphi')\ddot{x}_{To} + \sin(\theta')\cos(\varphi')\ddot{y}_{To} \end{bmatrix} \\
&= -m_B l R_2'^{-1} \begin{bmatrix} 0 \\ -\cos(\theta')\sin(\varphi')\ddot{x}_{To} - \sin(\theta')\sin(\varphi')\ddot{y}_{To} - \cos(\varphi')\ddot{z}_{To} \\ -\sin(\theta')\ddot{x}_{To} + \cos(\theta')\ddot{y}_{To} \end{bmatrix}
\end{aligned}
\tag{6.33}
$$

where

$$\ddot{\vec{r}}_{To} = [\ddot{x}_{To} \quad \ddot{y}_{To} \quad \ddot{z}_{To}]^T \tag{6.34}$$

Hence, the linear fictitious torques in body centred coordinates are

$$\tilde{M}_{FL} = -m_B l \begin{bmatrix} 0 \\ -\cos(\theta')\sin(\varphi')\ddot{x}_{To} - \sin(\theta')\sin(\varphi')\ddot{y}_{To} - \cos(\varphi')\ddot{z}_{To} \\ -\sin(\theta')\ddot{x}_{To} + \cos(\theta')\ddot{y}_{To} \end{bmatrix}$$
$$= -m_B Q(R_2'\ddot{\vec{r}}_{To}) = -m_B R_2 (Q_B R_2^{-1}(R_2'\ddot{\vec{r}}_{To})) \tag{6.35}$$

where

$$Q = \begin{bmatrix} 0 & 0 & 0 \\ 0 & 0 & -l \\ 0 & l & 0 \end{bmatrix}; \; Q_R = R_2^{-1}QR_2 \tag{6.36}$$

Observing that the acceleration of the tower axes relative to earth axes in blade body centred coordinates is $R_2'\ddot{\vec{r}}_{To}$ and in rotor axes coordinates is $R_2^{-1}(R_2'\ddot{\vec{r}}_{To})$, the linear fictitious torques in rotor axes coordinates are

$$M_{FL} = -m_B Q_B a_R \approx -m_B Q a_R \tag{6.37}$$

where $a_R = R_2^{-1}(R_2'\ddot{\vec{r}}_{To})$ and $Q_R \approx Q$, since both θ and φ are small.

The fictitious forces in Figure 6.15(b) and (6.13) are the sum of the angular fictitious torques, $-I_B \dot{\Omega}_R$, and the linear fictitious torques, $-m_B Q a_R$; that is,

$$\begin{bmatrix} M_{F\theta_R} \\ M_{F\varphi_R} \end{bmatrix} = -m_B l \begin{bmatrix} a_{Ry} \\ -a_{Rz} \end{bmatrix} - J \begin{bmatrix} \dot{\Omega}_{Rz} \\ \dot{\Omega}_{Ry} \end{bmatrix} \tag{6.38}$$

The accelerations in (6.38) are those that would be measured by an accelerometer in the hub. In the above development, gravity has been neglected. However, in the inertial reference frame dynamics, it need not be included. Since gravity is equivalent to a linear acceleration, it can be considered to be included in the linear acceleration components, a_{Ry} and a_{Rz}, incorporated into the fictitious torques; that is, the only requirement is that the measurement of linear accelerations should include the contribution due to gravity.

Utilising the fictitious torques (6.38) the control system for a single blade within the context of IBC is that shown in Figure 6.16. Since it is only dependent on the dynamics of the actuator and blade, design of the controller in Figure 6.16 is straightforward. This transparency of IBC together with the avoidance of information loss that occurs with the use of the Coleman transform enables the design of its controller to be more precise than IPC with more flexible and wider choice of objectives. For example, in Han (2012) the objective is the reduction

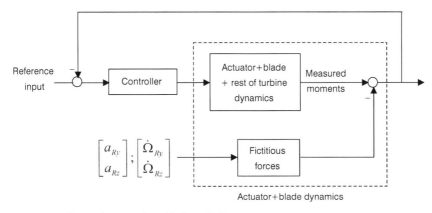

Figure 6.16 Control system for a blade in IBC

of the ultimate blade loads whilst in Han (2014) it is the reduction of both the DELs and ultimate blade loads. IBC is applied to the 5MW wind turbine. Since the greatest contribution to the blade and hub DELs is due to the 1Ω and 2Ω spectral peaks, the objective for the controller in Figure 6.16 is to reduce these. Three controllers are designed; the first, controller 1, has the objective to reduce the 1Ω peak in above rated wind speed; the second, controller 2, has the objective to reduce the 1Ω and 2Ω peaks in above rated wind speed; the third, controller 3, has the objective to reduce the 1Ω and 2Ω peaks but for a wind speed range extended below rated. The spectral density functions for the blade out-of-plane root bending moment and the hub torque about an axis in the rotor plane and rotating with the rotor are shown in Figure 6.17(a) and (b), respectively, for a mean wind speed of 22 m/s. The spectral density functions without IBC and with IBC, controller 1 and controller 2, are compared. Controller 1 effectively reduces the 1Ω spectral peak in both. Controller 2 effectively reduces the 1Ω and 2Ω spectral peaks in both and the 3Ω peak to some extent on the blade out-of-plane root bending moment. The reduction in blade and hub DELs is provided in Table 6.3.

Control schemes for improved farm operation

With the development of large offshore wind farms and attainment of high wind power penetration, it is no longer satisfactory for wind farms to be passive providers

Table 6.3 Blade bending moment the hub torque reductions with IBC

	Blade 1 M_x	Blade 1 M_y	Blade 2 M_x	Blade 2 M_y	Blade 3 M_x	Blade 3 M_y	Hub M_x	Hub M_y	Hub M_z
Controller 1	3.84	13.7	3.77	14.4	3.63	11.5	–0.41	23.9	24.3
Controller 2	3.61	15.8	3.65	17.1	3.47	14.3	–0.76	27.8	28.1
Controller 3	4.32	22	4.43	24	3.76	21.9	–0.74	49.3	45.0

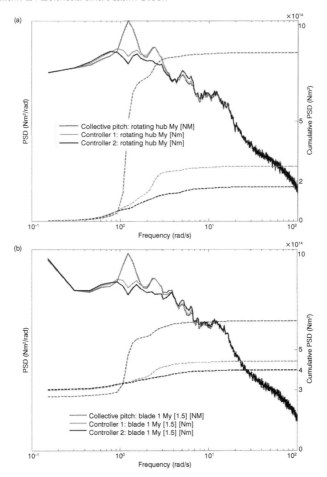

Figure 6.17 Spectral density functions for blade and hub with IBC

of generated power. Instead, offshore wind farms must become virtual generation plant that behave similarly to conventional generation. The power generated by the wind farm and its wind turbines can no longer simply be that dictated by the wind speed. The power must be adjusted as required by the operators. To do so requires flexible operation of the individual turbines and a wind farm controller to match power output to demand.

There are three possible points at which the power output from an offshore wind farm can be adjusted: at the onshore substation, at the offshore substation, or at the converter for each individual turbine, see Figure 6.18. Initiating the change in power at either of the substations would give rise to an imbalance between the aerodynamic power at the wind turbine and the electrical power transmitted. This imbalance must be quickly removed. Coordinated control action is, therefore, required at multiple points in the system. The approach preferred here is to make the power alteration at the wind turbines. The power

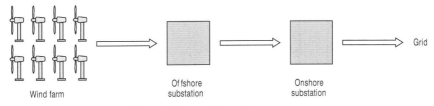

Figure 6.18 Connection of a wind farm to the grid

transmitted at the substations would naturally follow this change, requiring no additional modification to the system.

In addition to adjusting the power output, the wind farm controller could enable the wind farm to provide ancillary services such as curtailment, frequency support, voltage support, etc. Furthermore, there is extensive information regarding the local environment and conditions, including SCADA information, environmental information (wind direction, time of year, sea state etc., maintenance and repair logs information, wind farm layout information, condition monitoring and turbine health information) as well as individual wind turbine control information from nearby turbines. The potential to exploit this information through the wind farm controller to enable operators to make the most of their assets is substantial. In its most sophisticated form, the wind farm controller could control the operation of the individual wind turbines to achieve the most effective short- and long-term operation of all the assets in the wind farm. Accordingly, the general objectives for the wind farm controller is to maximise wind farm generated power, provide ancillary services, including curtailment, frequency support and voltage support, and minimise O&M costs.

Wind farm controller structure

A suitable hierarchical structure for the wind farm controller is depicted in Figure 6.19. The *network wind farm controller* acts in response to *network inputs*, i.e. inputs related to the grid side operation of the wind farm such as market information, power demand, grid frequency, etc. The *turbine wind farm controller* acts in response to *turbine inputs*, i.e. inputs related to the conditions and status of the N wind turbines in the farm. The network wind farm controller determines the total change in power output ΔP of the wind farm. For example, ΔP could be the difference between a demanded output P_0 and the actual output P_F. With a PI controller, zero steady state error would be achieved and so the power output would match the demand. When setting P_0, the network wind farm controller needs to take cognisance of the output corresponding to that dictated solely by the wind speed; that is,

$$P_W = P_F + \Delta P \qquad (6.39)$$

The *turbine wind farm controller* determines the allocation of the change in power to each turbine, i.e. the ΔP_i, $i = 1,\ldots,N$, such that

$$\Delta P = \sum_{i=1}^{N} \Delta P_i \qquad\qquad (6.40)$$

It receives information about the status of each turbine through the flags, S_i, $i = 1,...,N$.

By adopting the hierarchical structure in Figure 6.19 for the wind farm controller, it is possible to distribute a given change in power for the farm ΔP over the turbines in any manner, so long as the total change in power output is met. This has several advantages in comparison to simply distributing the change in power evenly between all the turbines. Firstly, the changes in turbine power can be more sensibly distributed across the farm. Secondly, if a wind turbine is unable to produce its requested change in power, say ΔP_i for turbine i, its contribution to the change in farm power output can be reallocated to other wind turbines. Thirdly, the structure is highly decentralised and is therefore easily scalable to very large wind farms. Finally, the approach presented here gives great autonomy to wind farm control designers.

The turbine wind farm controller does not introduce an additional power feedback round an individual turbine, since the only direct communication between it and the turbine are the status flags. There is an additional power outer feedback loop introduced by the network wind turbine controller but this is very weak. A simplified representation of this feedback loop is shown in Figure 6.20(a). The a_i, $i = 1,...,N$, are fractional allocations of the demanded change in power with

$$\sum_{r=1}^{N} \alpha_r = 1 \qquad\qquad (6.41)$$

The d_i, $i = 1,...,N$, represent the disturbances on each turbine induced by wind speed variations. In Figure 6.20(a), each turbine is depicted as a separate system, the G_i, $i = 1,...,N$. An equivalent system with all the wind turbines combined into a single system is depicted in Figure 6.20(b). Its open-loop transfer function $C \sum_{r=1}^{N} \alpha_r G_r$. The corresponding feedback loop for a single wind turbine, say turbine 1, is depicted in Figure 6.20(c). Its open-loop transfer

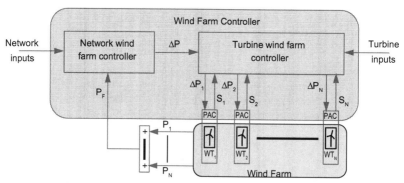

Figure 6.19 Wind farm control structure

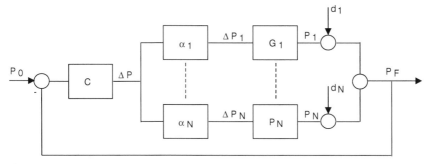

a) Simplified representation of network wind farm controller

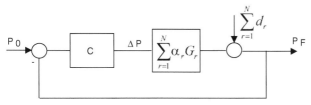

b) Feedback loop with turbines combined into single system

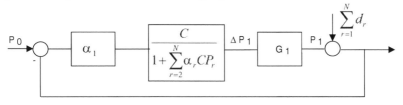

c) Feedback loop for single system

Figure 6.20 Network wind farm controller power feedback loop

function is $\alpha_1 CG_1 / (1 + C\sum_{r=2}^{N}\alpha_r G_r)$. When all the turbines are the same, with $G_1 = G_2 \cdots = G_N = G$, and equal allocation of the power adjustment is made, the open-loop transfer function for Figure 6.20(b) becomes CG and for Figure 6.20(c) becomes $\frac{1}{N}CG / (1 + \frac{N-1}{N}PG)$. Clearly, the outer power feedback loop through the network wind farm controller only acts as a very weak power feedback loop on a single turbine. Hence, as far as individual wind turbines are concerned, the wind farm controller does not introduce any additional significant feedback loops.

As in Figure 6.19, the wind farm controller allocates a demanded change in power output to each wind turbine. To realise this change, the wind turbine must be displaced from its normal operating conditions. In above rated wind speed, it is possible to do this by adjusting the torque demand with a simple increment ΔT. Because the full envelope controller regulates the rotor speed through pitch action in above rated operation, this change in torque gives rise to a disturbance in the rotor speed that is counteracted through pitch action without adjusting the generator torque. Hence, the power output is altered whilst the rotor speed is still controlled. However, in below rated wind speed, the full envelope controller regulates the rotor speed through adjusting the generator torque demand. The adjustment ΔT is treated as a disturbance and counteracted by the full envelope controller that uses generator reaction torque itself to regulate the turbine speed. An alternative approach in below rated wind speed is to either adjust the pitch angle $\Delta \beta$ or alter the full envelope controller operational strategy as necessary to provide the required change in power output.

Wind farm controllers with adjustments made to the individual turbine outputs using approaches similar to those discussed above have been investigated previously (Knudsen 2014). However, these do not provide a complete solution. Introducing a number of different operating strategies within the full envelope controller is complicated and requires detailed knowledge of the composition of the full envelope controller, whilst providing an increment to $\Delta \beta$ is relatively slow to respond and lacks accuracy and does not enable a temporary increase in power output in below rated operation. Additionally, if the method of operation differs in above and below rated wind speeds, then it may be difficult to achieve smooth switching.

The approach to adjusting individual turbine outputs presented in this section is in the form of an augmentation to a wind turbine full envelope controller. For this augmentation to be generic, it must fulfil the following requirements:

- The augmentation must be applicable to variable speed, pitch regulated machines without alteration to the turbine's full envelope controller.
- No knowledge of the design of the wind turbine's full envelope controller must be required.
- The augmentation must allow the operator to vary the power output of the wind turbine by an increment ΔP, however defined.
- The augmentation must enable the power output of the wind turbine to be altered quickly and accurately.
- The augmentation must smoothly accommodate switching between different modes of operation.
- The performance of the full envelope controller, including any gain scheduling, must not be compromised through the addition of the augmentation.

An augmentation to a wind turbine's controller that meets these requirements is the Power Adjusting Controller (PAC) (Stock 2015).

Power adjusting controller

In below rated wind speed, it is required that the power output is altered by adjusting the torque without altering the full envelope controller. Suppose that an increment ΔT is added to the torque demand from the full envelope controller and that an adjustment $\Delta\omega$ is added to the speed input to the full envelope controller, where $\Delta\omega$ is the change in speed caused by the increment in torque. The full envelope controller no longer counteracts the torque increment and so the power output is altered as required. There is now a discrepancy between the generator torque and the rotor torque and, depending on the sign of ΔT, the rotor either speeds up or slows down. However, the rotor speed can be brought back to its normal operating value by appropriate adjustment $\Delta\beta$ of the pitch angle. This adjustment must be made in such a way that the gain of the full envelope controller pitch control loop is not altered. The PAC is designed to act in the manner described above. It jackets the full envelope controller, see Figure 6.21. Because the power adjustment is made initially using an adjustment to the generator torque, the change in power can be fast limited only by the speed of response of the power electronics. The same method can be used in both above and below rated wind speed with no switching between modes of operation required. In addition, no knowledge of the design of the full envelope controller is required and so the controller could be implemented on any wind turbine, including retrofitting.

To ensure that the PAC does not compromise the performance of the full envelope controller, it must act as a feedforward modification to the full envelope controller and not introduce any additional feedback loops affecting its operation; that is, $\Delta\omega$, $\Delta\beta$ and ΔT_a can depend on ΔP but only very weakly on ω, β_d and T_d. The structure of the PAC is shown in Figure 6.22. The internal wind turbine models are very simple, depending only on a few basic parameters of the wind turbine, but include a dynamic inflow model. The introduction of additional feedback loops is avoided by exploiting aerodynamic separability (Leithead 1995, Jamieson 2011) and the design of the internal controller. Consequently, the PAC essentially dynamically adjusts the operating strategy of the wind turbine.

Through adjusting the operating strategy of the wind turbine, the PAC moves the operating point of the wind turbine away from the usual operating strategy.

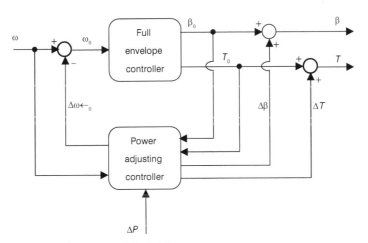

Figure 6.21 General structure of the PAC

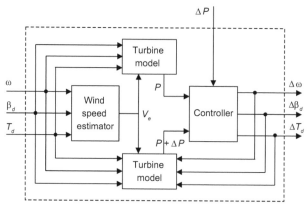

Figure 6.22 Structure of the PAC

Hence, it is important to ensure that the turbine remains in a safe operating region. Indeed, in some circumstances, it may not be possible to obtain a pitch angle that balances the aerodynamic and generator torques, e.g. when a positive ΔP is requested in below rated operation. The rotor then either slows down or speeds up unless provision of the requested ΔP is curtailed. To prevent operation outside of the safe operating region, limits are applied to the operation of the PAC. They include minimum and maximum torque bounds, which may vary with generator speed, and generator speed bounds. These limits are referred to as the PAC supervisory rules. When limits are reached or in some cases breached, flags are set within the PAC. A summary of the PAC supervisory rules (Stock 2015) is provided below.

The occurrence of events triggered by the supervisory rules is communicated between the PAC and wind farm controller using flags. S_i, see Figure 6.19, residing in the PAC. (Capital letters are used to indicate flag names with sub-flags in bracketed italics.) There are two sets of rules, black rules defined by a boundary on the torque/speed plane that act as a hard limit and traffic light rules, defined by two concentric boundaries contained within the black rules boundary, that act as soft limits. Maximum aerodynamic and drive-train torque boundaries apply. The regions inside the inner traffic light boundary, between the inner and outer traffic light boundaries and outside the outer traffic light boundary are designated green, amber and red, respectively.

General supervisory rules:

- The requested change in power, rate of change in power and pitch rates are subject to limits and the permissible turbulence intensity and wind speed are subject to upper and lower limits, respectively. These limits and events designated high priority, e.g. requests for synthetic inertia, are defined with agreement and cannot be changed without agreement of the OEM.
- The PAC is turned on when the PAC ON flag is set at a request from the wind farm controller.

- The PAC is turned off when the PAC ON flag is reset by either the PAC itself or by the PAC at a request from the wind farm controller. The PAC goes into recovery mode and the RECOVERY flag is set. The speed of recovery is fast or slow depending on the setting of the RECOVERY (Fast/Slow) flag and sub-flag. The sub-flag (Fast/Slow) can be reset at the request of the wind farm controller. The default setting is RECOVERY (Fast). During the recovery mode the PAC rejects any requested change in power. The REJECTION (Recovery) flag is set by the PAC. On completion of recovery mode the RECOVERY (Complete) flag and sub-flag are set and the PAC ON flag is reset.
- Only black supervisory rules apply to high priority events. The PRIORITY flag is set by the PAC at a request from the wind farm controller.
- If the limit for requested change in power is exceeded, the REJECTION (Power) flag is set by the PAC.
- If the limit for requested change of power rate limit is exceeded and the PRIORITY flag is not set, the rate limit applies and the REJECTION (Power rate) flag is set by the PAC.
- If the turbulence intensity limit is exceeded, the PAC ON flag is reset and latched and the PAC ON (Turbulence) sub-flag is set and latched by the PAC.
- If the actuator pitch rate limits are violated by the turbine full envelope controller, the PAC ON flag is reset indefinitely and the PAC ON (Actuator) sub-flag is set indefinitely by the PAC.
- If the low wind speed limit is exceeded, the PAC ON flag is reset and latched and the PAC ON (Wind Speed) sub-flag is set and latched by the PAC.
- If the turbine state is divergent such that normal operation is unreachable, the DIVERGENT flag is set by the PAC.

Black supervisory rules:

- The boundary and maximum possible generator reaction torque are set with agreement and cannot be changed without agreement of the OEM.
- The boundary should not be crossed under any circumstances. If the turbine state is outside the boundary, the PAC ON flag is reset by the PAC.
- On the turbine state reaching the boundary, the REJECTION (Limit) flag and sub-flag are set by the PAC.
- If the turbine state remains on the boundary beyond a pre-set time limit, the PAC ON flag is reset by the PAC.
- On a section of the boundary corresponding to the maximum possible generator reaction torque, the permitted time limit before resetting the PAC ON flag is zero.

Traffic light supervisory rules:

- The boundaries can be set at a request from wind farm controller.
- The maximum magnitude of change of power in all regions can be set by the wind farm controller subject to the fixed upper limit, the maximum

magnitude for the amber region being less than the maximum for the green region and the maximum/minimum change of power for that part of the red region to the left/right of the operating strategy being zero.

• When the turbine state is in the green/amber/red region, the corresponding GREEN/AMBER/RED flag is set by the PAC.

• When the demanded change in power exceeds the maximum or minimum, the corresponding REJECTION (Green Limit)/(Amber Limit)/(Red Limit) flag and sub-flag are set by the PAC.

The PAC consistently provides an accurate change in power output without causing large changes to the generator speed, except in the case of an increase in power output, when a large change in the generator speed is unavoidable due to there being no pitch angle available that can balance the aerodynamic and generator torques. The performance of the PAC, when implemented on a Simulink model of the 5MW wind turbine, is illustrated by Figures 6.23 and 6.24. The results are from a simulation for a version of the PAC that did not include dynamic inflow contributions to the internal turbine models. When a reduction in power output is requested, the generator speed initially increases. The pitch angle is subsequently increased to minimise the change in generator speed. When an increase in power is requested, e.g. between 170 and 200 seconds in Figures 6.23 and 6.24, the generator speed necessarily reduces as the aerodynamic and generator torques cannot be equalised. In all cases, the actual change in power tracks the requested change in power very well. The results presented here are typical for all requested changes in power, and for all wind speeds simulated.

The performance for a requested 100 kW step change in power of the PAC, when implemented on Bladed simulations, demonstrated the need to include dynamic inflow. For example, with the 1.5MW wind turbine without the inclusion of dynamic inflow in the PAC, the initial response is under-damped with an overshoot of as much as 80 per cent and a large oscillation with period approximately 30 seconds. The performance with dynamic inflow included in the PAC is illustrated by Figure 6.25. It is possible to tune the representation of dynamic inflow in the PAC. Bladed simulation results with both a tuned and unturned dynamic inflow representation are shown in Figure 6.25.

Application of PAC to synthetic inertia and droop control

The reason for developing the PAC is to embed it into a wind farm controller with the hierarchal and decentralised structure in Figure 6.18 with the role of adjusting the power output from each turbine in the farm by the amount requested by the wind farm controller. Of course, the combined action of the network wind farm controller, the turbine wind farm controller and the PACs must meet the requirements of the wind farm. In particular, the wind farm controller must be stable and match actual output from the wind farm to demand accurately and rapidly when necessary. In addition, through the use of the status flags, the allocation that each turbine makes to the demanded output of the wind farm

a) Change in power

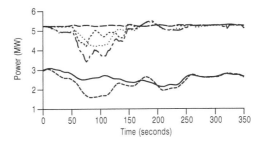

b) Change in generator speed

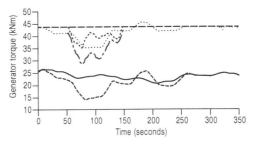

c) Change in generator torque

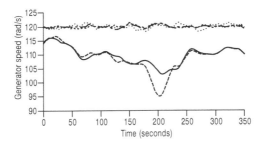

c) Change in pitch angle

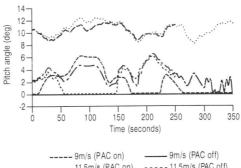

----- 9m/s (PAC on)   ——— 9m/s (PAC off)
— - - 11.5m/s (PAC on)   - - - - - 11.5m/s (PAC off)
.... 15m/s (PAC on)   – – – 15m/s (PAC off)

Figure 6.23 Application of PAC to 5MW wind turbine Simulink mode

a) Change in power

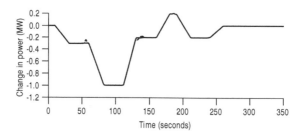

b) Change in generator speed

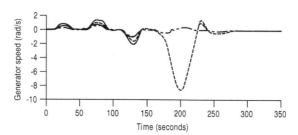

c) Change in generator torque

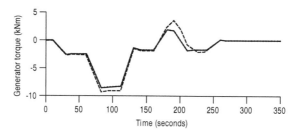

d) Change in pitch angle

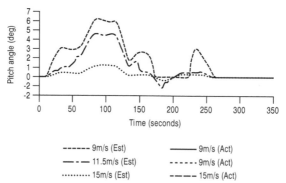

------ 9m/s (Est)   ——— 9m/s (Act)

— · — 11.5m/s (Est)   ----- 9m/s (Act)

··········· 15m/s (Est)   ----- 15m/s (Act)

Figure 6.24 Application of PAC to 5MW wind turbine Simulink model – change in variables

a) Change in power

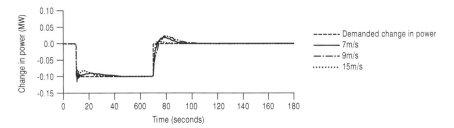

b) Change in generator speed

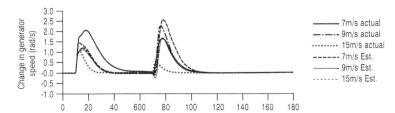

c) Change in generator torque

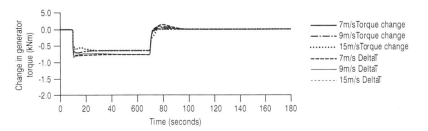

d) Change in pitch angle

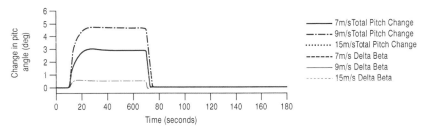

Figure 6.25 Application of PAC to 1.5MW turbine Bladed – change in variables

must not cause the turbine to leave its safe operating region. That the wind farm controller with the PACs embedded does so is explored in the context of power curtailment in Hur (2014).

Whether within the context of a wind farm or a single wind turbine, the PAC can be used to provide grid frequency support in the form of either synthetic inertia (Stock 2012) or droop control (Stock 2014). Both are discussed below.

Synthetic inertia is the provision of an increase in power proportional to the rate of change of grid frequency in order to mimic the inertial response of conventional synchronous plant.

Conventional synchronous plant have an inertia constant, H, of approximately 6 secs. The inertia constant is related to the rate of change of frequency by

$$S\frac{df}{dt} = \frac{\Delta Pf}{2HS} \tag{6.42}$$

where ΔP is the change in power output, f is the grid frequency, and S is the rated power. For a wind turbine to produce an equivalent inertial response, the change in power for a given change in frequency must be proportional to the rate of change of frequency (assuming the total change in frequency is small) with scaling constant

$$K_{inertia} = -\Delta P / \frac{df}{dt} = -2SH / f_{nominal} \tag{6.43}$$

When the grid frequency is recovering from a frequency drop the rate of change of frequency is positive. To realise synthetic inertia, the required change in generated power is simply input to the PAC. Because providing a reduction in power when the frequency is returning towards the nominal value is undesirable, the change in power output requested by synthetic inertia is bounded below by zero. A maximum value is not required, since the PAC supervisory rules already limit the maximum requested change in power.

Bladed simulations indicate that synthetic inertia equivalent to an H value of 18 secs can be safely provide by the PAC even when the frequency drop is large and the wind speed low. For the 5MW wind turbine providing synthetic inertia with an H value of 18 secs through the PAC, a grid frequency event as shown in Figure 6.26 is simulated. A measured frequency event is, also, shown in Figure 6.26 for comparison. The resulting power and change in power are shown in Figure 6.27 and the torque is plotted against rotor speed in Figure 6.28. The boundaries for the *green*, *amber* and *red* regions defined in the PAC rules are also depicted in Figure 6.28. It can be seen that the operating point of the wind turbine remains well within the *green* safe operating region. The ability to provide synthetic inertia that is greater than the inertia that is typically provided by a conventional synchronous generator, opens up the possibility of using wind farm control to distribute the provision of synthetic inertia unevenly over the wind turbines in the farm. Hence, the response of the farm could be equivalent to a conventional synchronous generator without needing to use all the wind turbines in the farm.

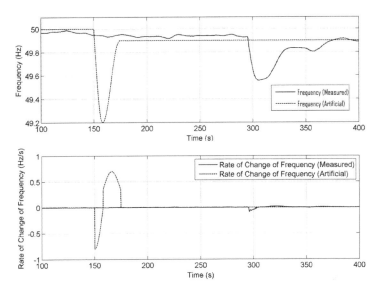

Figure 6.26 Grid frequency inputs to synthetic inertia

Droop control is the provision of a change in power proportional to the change in grid. Whilst synthetic inertia reduces the rate of change of grid frequency, droop control is used to reduce the drop in grid frequency through a power increase proportional to the deviation in grid frequency from the set value (typically 50 Hz). To supply droop control, increases in power are required that may be sustained for prolonged periods of time and so the turbine must operate with a reduced power output in order to provide headroom for power increases.

The UK grid code is fairly typical of grid codes for modern industrialised countries. It requires synchronous generation to have a droop capability of 3–5 per cent. This implies that a change in frequency of 3–5 per cent would cause a change in power output of 100 per cent. It is also a requirement that the frequency should be kept between 49.8 and 50.2 Hz and so the maximum change in power output required during normal operation is about 10 per cent of output power. The required change in output power ΔP is directly related to the grid frequency; that is,

$$\Delta P = \Delta P_{offffset} + \Delta P_{freq} = \Delta P_{offset} + K_{freq}\left(f_{nominal} - f\right) \tag{6.44}$$

where ΔP_{offset} is the power offset set according to Figure 6.29 for the 1.5MW and 5MW wind turbines and K_{freq} is a constant set such that $P_{freq} = \Delta P_{offset}$ at the maximum allowable grid frequency. To realise droop control, the required change in generated power is simply input to the PAC.

Strategy 1 in Figures 6.29(a) and (b) provides droop control proportional to the rated power output; that is, the response to a given change in grid frequency is the same regardless of the wind speed at the rotor. Strategy 2 in Figures 6.29(c) and (d) provides a greater change in power output at higher wind speeds. In low

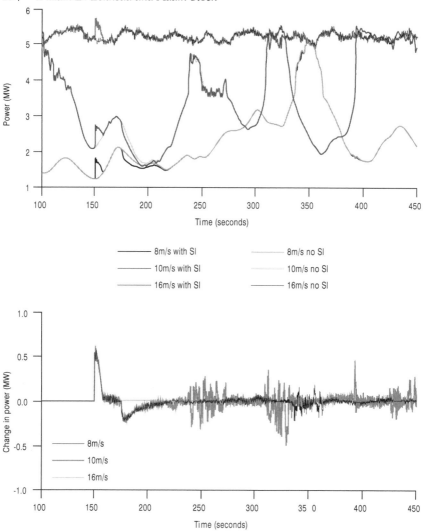

Figure 6.27 Power and change in power for the 5MW wind turbine providing synthetic inertia

wind speed, the offset is zero to turn off the PAC and, in wind speed significantly above rated, the offset is again zero, since the power output can be increased indefinitely by adjusting the pitch to increase aerodynamic torque. At the two transition wind speeds, hysteresis is used in both strategies with the red offset in Figure 6.29 applying to increasing wind speed and the black offset applying to decreasing wind speed.

The performance when delivering droop control through the PAC is assessed by aggregating the response of 18 wind turbines. Droop control is applied to each turbine individually, although it is possible to use wind farm control to distribute

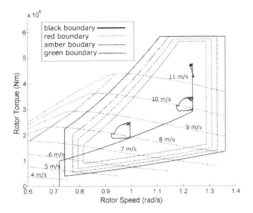

Figure 6.28 Torque versus rotor speed for the 5MW wind turbine providing synthetic inertia

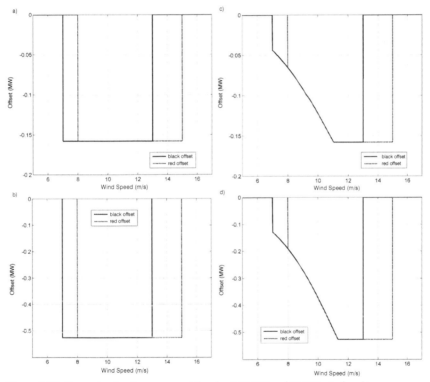

Figure 6.29 Strategies for droop control for the 1.5MW and 5MW wind turbines

the required change in power for droop control over the wind turbines in a farm in a more advanced manner. Typical Bladed simulation results with six 5MW turbines having a mean wind speed of 8 m/s (below rated), six having a mean wind speed of 10 m/s (close to rated) and six having a mean wind speed of 14 m/s (above rated) variable wind speed are shown in Figure 6.30. All the turbines are following strategy 2. The frequency input used is measured data from the grid.

a) Average power for each group of 6 turbines with and without the   PA C

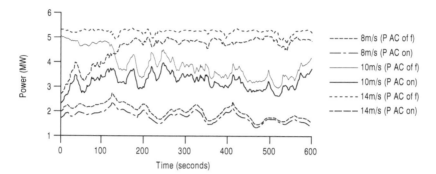

b) Average actual and requested change in power for each group

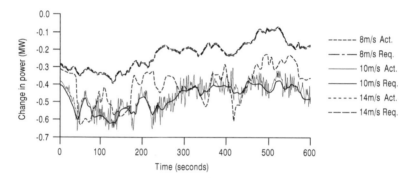

c) Average demanded and actual change in power for all 18 turbines

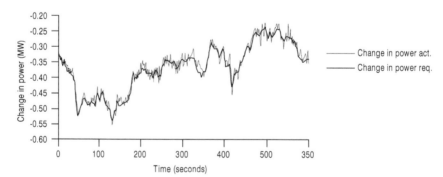

Figure 6.30 Droop control with 18 5MW wind turbines (strategy 2)

The simulations demonstrate that the PAC is capable of providing accurate droop control. Operating wind turbines with droop control capabilities necessarily reduces their energy capture. Estimates of the reduction in power output over the lifetime of the turbine are between 3.1 per cent and 4.7 per cent depending upon the strategy used, the wind turbine used, and the wind conditions at the site.

All performance assessments of the PAC indicate that there is no significant increase in either the ultimate loads or the lifetime damage equivalent loads. Indeed, for the latter, a reduction is observed for many loads but only about 1 per cent.

References

Bossanyi, E.A. (2003a). Wind turbine control for load reduction, *Wind Energy*, V6, 2003.

Bossanyi, E.A. (2003b). Individual blade pitch control for load reduction, *Wind Energy*, V6, 2003.

Bossanyi, E.A. (2005). Further load reduction with individual pitch control, *Wind Energy*, V8, 2005.

Chatzopoulos, A., Leithead W.E. (2010a). Reducing tower fatigue loads by a coordinated control of the SUPERGEN 2MW exemplar wind turbine, Torque2010, Heraklion, June 2010.

Chatzopoulos, A. (2010b), Full envelope wind turbine controller design for power regulation and tower load reduction, PhD Thesis, University of Strathclyde, 2010.

Chatzopoulos, A., Leithead W.E. (2011), The use of a novel power coordinated controller for the effective reduction of wind turbine tower fatigue loads, ERA6, Piraeus, September 2011.

Feynman, R.P., Leighton, R.B., Sands, M.L. (2006). *The Feynman Lecture on Physics*, Vol 1, Pearson/Addison Wesley, San Francisco, 2006.

Han Yi, Leithead, W.E. (2012). Alleviation of extreme blade loads by individual blade control during normal wind turbine operation, EWEC2012, Copenhagen, Denmark, April, 2012.

Han Yi, Leithead, W.E. (2014). Combined wind turbine fatigue and ultimate load reduction by individual blade control, Torque2014, Copenhagen, June 2014.

Hur, S.H., Leithead, W.E. (2014). Curtailment of wind farm power output through flexible turbine operation using wind farm control, EWEA2014, Barcelona, Spain, 2014

Jamieson, P., Leithead, W.E., Gala-Santos, M.S. (2011). The aerodynamic basis of a torque separability property, EWEA2011, Brussels, Belgium, 2011.

Knudsen, T., Bak, T., Svenstrup, M. (2014). Survey of wind farm control—power and fatigue optimization, *Wind Energy*, 2014.

Leith, D., Leithead, W.E. (1996). Appropriate realisation of gain-scheduling controllers with application to wind turbine regulation, *IJC*, V73 No.11, 1996.

Leithead, W.E., Rogers, M.C.M., Leith, D.J., Connor, B. (1995). Design of wind turbine controllers, Proceedings of EURACO Workshop, Recent results in robust & adaptive control, Florence, Italy, 1995.

Leithead, W.E., Dominguez, S. (2005). *Controller Design for the Cancellation of the Tower Fore-Aft Mode in a Wind Turbine*, Proc. CDC/ECC2005, Seville, Spain, December 2005.

Leithead, W.E., Dominguez, S. (2006) Coordinated control design for wind turbine control systems, EWEC2006, Athens, March 2006.

Leithead, W.E., Neilson, V., Dominguez, S. (2009a). Alleviation of unbalanced rotor loads by single blade controllers, EWEC2009, Marseilles, France, March 2009.

Leithead, W.E., Dominguez, S., Gonzalez, C., Dutka, A. (2009b). Full load alleviation via the intelligent actuator, AWEA2009, Chicago, USA, May 2009.

Leithead, W.E., Neilson, V., Dominguez, S. (2009c). A novel approach to structural load control using intelligent actuators. Proc.17th Med. Control and Automation Conf., Thessaloniki, Greece, June 2009.

Neilson, V.W. (2010). Individual blade control for fatigue load reduction of large-scale wind turbines: theory and modelling, MPhil Thesis, University of Strathclyde, 2010.

Stock, A., Leithead, W.E. (2012). Providing grid frequency support using variable speed wind turbines with augmented control, EWEA2012, Copenhagen, Denmark, 2012.

Stock, A., Leithead, W.E. (2014). Providing frequency droop control using variable speed wind turbines with augmented control, EWEA2014, Barcelona, Spain, 2014.

Stock, A. (2015). Augmented control for flexible operation of wind turbines, PhD Thesis, University of Strathclyde, 2015.

7 Assessment of operation and maintenance strategies

Simon Hogg, Behzad Kazemtabrizi and Richard Williams

Owner/operator requirements

Introduction

Significant future growth in UK wind power capacity will be almost exclusively offshore. The next generation of UK offshore wind farms are moving further offshore, becoming larger in both capacity and physical size, and are being positioned in deeper water than has been the case for many of the previous projects. Significant investment and planning is already underway to achieve this goal but this needs to be accelerated and costs reduced in order to ensure the long-term aim for the deployment of large-scale offshore wind farms is realised. Economies of scale will contribute to bringing down the cost of energy from offshore wind, but additional reductions in the way that wind farm assets are managed (installation, operations and maintenance costs) are also needed in order to finally free the industry from its reliance on heavy financial subsidies. The challenges are vast and encompass the installation phase, operational climate, supply chain, manufacture, installation and operations and maintenance activity (O&M).

This section focuses on an assessment of the current industrial view concerning O&M of large far-offshore wind farm assets. The data used to assess owner/ operator O&M requirements was collected from interviews carried out with a wide range of stakeholders from across the wind industry.

The interviews showed that at present, lessons that have been learnt from previous generation smaller-scale projects are not always passed on and adopted by operators of new farms as they come online and the industry grows. The interviews also identified that a higher degree of product standardisation for the new large-scale projects is mandatory for cost reduction to safeguard the future of the industry. Beyond the technical challenges of construction and innovation, there is a need for greater development of operational strategies as wind farms increase in size, essentially becoming off-shore wind power stations that will need to be able to compete on an equal footing with more conventional generating plant.

Figure 7.1 shows a breakdown of the various contributions to the cost of energy (total = £140/MWh) for a typical large (4MW) offshore wind turbine installed in UK waters, based on 2011 data. It can been seen that operating costs including transmission charges account for approximately one third of the cost of

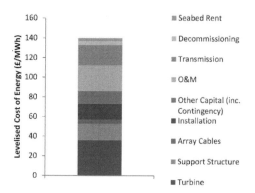

Figure 7.1 2011 levelised cost of energy breakdown for a 4 MW-class offshore turbine (Data from 'Offshore Wind Cost Reduction Pathways Study', Crown Estate, 2012.)

generating energy from these assets. Typically O&M costs contribute between 10 and 35 per cent of the cost of energy from currently operating offshore wind assets depending upon location, technology, and age. Reducing O&M costs through improved operational strategies has significant potential for reducing the cost of energy from offshore wind.

Operational strategies must also aim to maximise the wind farm capacity factor by increasing the energy based availability, to maximise the revenues generated from wind farms. A component failure can significantly affect energy production especially during high wind periods. Failure of certain components, such as the transmission cable, can be catastrophic in terms of energy output as they affect the whole farm. Other types of failure, e.g. an unexpected gearbox failure, may only have a localised impact on energy generation operations at the farm level but can still have significant maintenance cost implications.

Literature review

There is already a significant literature base on the management of offshore wind farm assets. Some of the important contributions to the literature are reviewed in this section. The outcome from these studies was used to benchmark the information gathered from the interviews carried out with current industry practitioners that is described later in this chapter. Ultimately, a combination of information gathered from the literature and from the interviews, has enabled conclusions to be drawn concerning knowledge gaps and identification of key areas for future research.

El-Thalji and Liyanage (El-Thalji 2012) presented a systemic review of wind turbine maintenance O&M practices in 2012, following their earlier paper in 2010 (El-Thalji 2010). Overall they found a significant breadth of O&M literature. The main academic contributions included condition monitoring, diagnostic and prognostic techniques that are largely developed from already established technologies used in other industries. Some modelling of maintenance practices has also been undertaken and this has been reviewed by El-Thalji and Liyanage (El-Thalji 2010) and Alsyouf and El-Thalji (Alsyouf 2008). In the latter study it was concluded that there is an ongoing need for research to look at how to apply maintenance practices that have already been developed for other industries into the wind industry. Furthermore, they highlighted the need for closer collaboration between wind-industry stakeholders to improve O&M technologies and methodologies, including between the original equipment manufacturers (OEMs) and the owners and operators.

A significant number of studies on wind turbine and wind farm reliability have already been published in the open literature, based on the limited amount of reliability data that has been made available to workers by the OEMs and wind farm operators. Many references (Crabtree 2012, Wilkinson 2011, Lange 2010, Spinato 2008, Ribrant 2007, Tavner 2007) are examples of failure rate analysis studies using this data. Much of the work uses data from wind system database repositories. Hameed (2011) argued for a reliability, availability, maintainability and safety database, that will allow information to be shared across the wind industry in a similar manner to systems that are available in the oil and gas industry. The Offshore Renewable Energy Catapult Centre is starting to move in this direction with their SPARTA project, although at the moment the database will only be available to contributing industry partners. A significant amount of data was collected and analysed under the Reliawind project (Wilkinson 2010, 2011, Lange 2010, Spinato 2008, Ribrant 2007, Tavner 2007, Hameed 2011), in order to quantify failures and their mechanisms. Most of this data corresponds to onshore turbines as little data is available to academia from offshore wind farms, in particular far-offshore wind farms for which essentially no data is available at present.

The analysis carried out so far on the limited offshore data available has resulted in some interesting observations. For example, Crabtree (2012) studied correlations between operational condition monitoring data and wind speed. In some cases up to 40 per cent of the potential energy available was being lost due to unavailability of turbines. Failures were occurring at higher wind speeds and repairs were delayed by reduced accessibility due to high wind and wave conditions and the reactive maintenance strategies employed. Harman (2008) showed that availability increased as the farm matured, a result of the elimination of early teething problems and operating procedures being improved. Only 10 years of data were available in Harman's study and so it was not possible to confirm this trend over the full operating life of the turbine assets. Similarly to the trends observed by Crabtree (2012), the availability was found to reduce with increased wind speed, due to its impact on access for repairs. It was also observed that at low wind speeds, turbine availability is impacted by the execution of planned maintenance interventions, which have to be carried out under favourable weather conditions.

Many operation and maintenance strategies for wind farms have been proposed and several reviews of these exist (e.g. Alsyouf 2008, El-Thalji 2010). The practices have evolved from onshore methodologies (Andrawus 2006) to the offshore environment (e.g. McMillan 2007, Besnard 2009, 2010a, Wilkinson 2007, Amirat 2009, Hameed 2009, Nilsson 2007).

Few studies exist for far-offshore turbine arrays. Utne (2010) investigated the development of offshore maintenance strategies, planning and execution. The conclusion from this study was that a proactive strategy with a focus on condition-based preventative maintenance was essential for minimising the cost of energy from offshore assets. The challenge of cost optimisation was described as being complex due to the number of diverse stakeholders involved. Managing the interfaces and the flow of information across them was found to be a major element of cost optimisation, and the approach described in the study begins to address this. Rademakers (2003) suggested that preventative maintenance strategies compared with reactive ones possess the potential to reduce maintenance costs by up to a factor of 2 for offshore farms, compared with onshore farms where these two strategic approaches result in similar maintenance costs over the life of wind farms. It can be assumed that the offshore cost benefit factor of 2 will increase even further for far-offshore farms, where the more challenging environment makes careful planning of maintenance activities even more essential. McMillan (2007) showed that the use of condition monitoring onshore is only just economically viable. When unplanned outages increase, the economic benefits from condition monitoring increase significantly, justifying the use of condition monitoring systems for offshore farms. Several risk-based methods are available in the literature (Sørensen 2007, Nielson 2010) for analysing the operation and maintenance of offshore wind farms.

Several studies have been published on cost modelling of offshore O&M scenarios (Eecen 2007, Byon 2010a, 2010b). Van de Pieterman (2011) modelled a condition based maintenance strategy that employed workboats operating from a mother ship. It was found that the long waiting times for suitable weather windows, which resulted in expensive equipment lying idle, had a big impact on overall costs. Besnard (2010b, 2009) proposed a model based on opportunistic maintenance activity which took advantage of periods of low wind speed, when access is easiest and the cost of lost energy generation is low. The model was run on a day-by-day basis and maintenance tasks were planned in an optimised manner on each day. Maintenance cost reductions of up to 45 per cent were calculated for the test cases used in the study.

The use of condition monitoring within wind systems is increasing as the industry matures and codes and standards now require that at least some form of condition monitoring is installed on new wind turbines. Traditionally, condition monitoring on wind turbines has been used to flag alarms which result in operators reactively inspecting the asset. Reliability-based maintenance uses condition monitoring systems or visual inspections to assess the severity of damage to a system. Two examples of condition monitoring simulations applied in the wind industry are Besnard (2009) and Marseguerra (2002). Modelling a system first requires that a predictive model of the systems components is established. The threshold for the condition of each system component when action needs to be taken must then be established. This type of model can be used to optimise maintenance activities for minimum cost.

Health monitoring is widely used in the management of wind farm assets (McMillan 2007, Amirat 2009, Hameed 2009, Watson 2006, McMillan 2008,

Wilkinson 2007). Supervisory Control Data Acquisition (SCADA) and condition monitoring systems are used to measure and record the condition of turbine components and the turbine's operating environment. This information is used to predict, detect and diagnose faults as they emerge. Amirat (2009), Hameed (2009) and Lu (2009) are some of the reviews of condition monitoring fault diagnostic techniques that have been published to date. The study by Lu focuses principally on the turbine drive train components, whereas Hameed carried out a broader review of the complete wind turbine system and the full range of sensors and monitoring techniques used. Amirat's study looked at the detection of drive train faults from electrical signals in the output from the generator. Some of the more advanced techniques are relatively expensive, however the costs can be justified for offshore turbines, where more sophisticated condition monitoring compared with onshore turbines is needed to reduce the risk associated with unplanned reactive maintenance in the offshore environment (Utne 2010). The cost of condition monitoring systems for offshore turbines is also influenced by the need for increased reliability and hence greater redundancy.

Studies have also been carried out to model wave height and wind condition in order to predict access to offshore turbines for maintenance interventions. Access is widely recognised to be one of the more important contributors to the cost of energy from offshore wind farms. The studies have shown that the availability of access to turbines for maintenance activities is site specific. Experimental data usually requiring the installation of a met-mast is needed for validation purposes for each site and so the costs of acquiring this data are high. Van Bussel (2003) investigated the transport of maintenance crews for a wind farm 43 km offshore and concluded that for the particular site studied access availability of above 90 per cent could be achieved using a flexible gangway system capable of operating in wave heights of up to 3 m. This study showed that a hybrid O&M strategy using different types of access at different times of year was favourable. The work did not, however, include significant failures requiring jack-up vessels or winch operations.

Survey of owner/operator requirements

The SUPERGEN Wind project team carried out a survey of the various stakeholder groups in order to map owner/operator requirements. The methodology used to carry out the study and the results obtained are described in this section.

The survey was carried out by holding interviews with representatives from a range of organisations with interests in the wind turbine industry. All of the participants were either directly involved in current offshore wind operations, or they were involved in the development of operations and maintenance planning for future large offshore wind farm installations. The participating organisations included:

- Offshore and onshore developers
- Current offshore operators
- Technical wind energy consultants
- Infrastructure developers.

Several of the organisations surveyed were onshore wind farm operators who were also engaged in the development of future large UK Round 3 offshore sites. This was an important group for the survey as it enabled an assessment to be made of the main concerns and challenges facing those companies who have a history of onshore operating experience with little or no direct offshore experience, but who are currently moving towards significant deployment of offshore assets.

Interviews were conducted by asking questions from a standard template. The organisations contacted for the survey included PB Power, Göteborg Energy, National Grid, DONG Energy, EDPR, DNV and OREC.

Questions were asked in the survey to explicitly solicit stakeholder views on: the functions involved in managing assets and their inter-relationships, the challenges associated with access, O&M strategy and prioritisation and health monitoring of assets in the development of O&M technology. The outcomes from the survey in each of these areas is summarised in the following sub-sections.

Functions involved in managing offshore wind farm assets and their inter-relationships

The major functions involved in the O&M management of wind farm assets were identified to be:

- Management by owner
- Operations management
- Health monitoring
- Maintenance management
- Field maintenance
- Information management
- Data repository and management.

Figure 7.2 shows an interface map of the lines of communication between the functions. All of the functions are inter-related, but in many cases the connection is not direct. For example, the owner's main focus is on the return

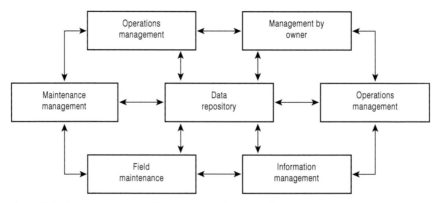

Figure 7.2 Asset management functions and their interfaces

on their investment which is clearly highly dependent upon an appropriate and ideally optimised asset management strategy. The owners do not need to be directly involved in the day-to-day management of field maintenance activities, but they do need to be convinced the operations management has the correct maintenance management structures in place.

Issues surrounding confidentiality of company data and practices was identified in the survey as a major operational concern. The offshore wind industry is still relatively young and organisations that have made significant investment have a strong focus on protecting their intellectual property and operational data in what is a highly competitive environment. This was highlighted in the discussions with most of the stakeholders. This can often impact the end of the warranty period, where control of assets transfers from the OEM to the owner/operator, sometimes with little sharing of historical data and information about the operational life of the turbine between the two parties. This can be a particular concern when, for example, the OEM may have replaced a significant component without informing the operator. As a result, the operator can miss potentially important information on component failures which can affect their other turbine assets. Possible solutions suggested so far include forming joint ventures between OEMs and operators during warranty periods, so that the early operation is transparent to the operator whilst the support of the OEM is maintained. Better sharing of information is considered to be mandatory for optimising the efficiency and minimising the cost of energy generation from wind farm assets. Onshore experience is that these relationships between OEMs and operators have generally improved with time and it is anticipated that this learning will be transferred to offshore installations. However, more structured and planned approaches are still needed in this area.

At the time that the survey was carried out few of the UK's large offshore wind farms were out of their warranty periods. In the majority of cases O&M operations were being undertaken by OEMs and information on the performance of assets was rarely being shared with operators. There was, however, some evidence of change within the industry. Examples were identified in the surveys where some of the larger operators had agreed greater access to warranty period data with OEMs and were now able to have a stronger influence on O&M decisions made during the early life of their wind farm assets.

Access challenges

Access to offshore wind assets was identified as being a major challenge for both optimising the cost of O&M operations and ensuring a safe working environment. Access can be divided into two fundamental operations, travel to the location of the asset and transfer onto it. Logistics planning for the movement of boats, equipment, spares and personnel was identified as one important area where improvements are needed to optimise the effectiveness of operations. Better systems for transferring from vessels onto offshore assets in unfavourable weather and sea-state conditions to open up access time windows, was highlighted as an area where new developments could have a significant impact on reducing the cost of energy and improving the safety of offshore operations. The importance of taking care to safeguard the environment for birds, fish and other wildlife and minimise pollution when planning O&M activities so that farms operate in a low-impact as well as low-carbon manner, was also highlighted in the study.

The maintenance practices adopted in the wind industry have been largely developed onshore. Usually this involves conducting inspection visits to the assets to identify and diagnose emerging problems that have been detected by health monitoring systems, before a repair or other maintenance action is performed often during a separate follow-up visit. This practice has served the industry well onshore where access rarely presents any significant challenges. Vehicle access to the base of turbine towers is almost always possible onshore and there are few limits on access imposed by weather with the exception of wind speed restrictions on hub or nacelle access. Similar approaches have been used for first generation near-coast offshore sites. Access using relatively small service boats is common, but the move to sites further offshore challenges these practices. Harsher wind and wave environmental conditions place a much greater restriction on access opportunities and have a significant impact on the availability of O&M working days. The experience reported in the survey indicated that even at relatively short distances from shore, weather window restrictions can reduce turbine accessibility to around 50 per cent of time.

For many planned far-offshore sites the distances are too great for land-based O&M service centres to be a viable option. Some turbines planned for the UK's Round 3 sites are beyond 100 km from shore. Boat access from land would take many hours and the opportunity to carry out any significant maintenance activity with the timeframe of a single day is, in many cases, non-existent. There are two principal access strategies for overcoming this.

The first of these involves the deployment of offshore O&M bases either using fixed or floating structures located close to or within the wind farm. Floating bases clearly have the advantage of being relatively easily relocated which might bring additional benefits for long-term maintenance planning. Personnel would remain offshore for periods of time in a similar manner to current practices in the oil and gas industry. Turbines would be accessed by boat from the offshore base using vessels similar to those already in use for near-shore operations. Sea-state and wind restrictions would remain, but with greatly reduced weather window durations needed to carry out maintenance and so access availability would be greatly improved, in addition to the increase in the speed of operations.

Use of helicopters for access to offshore turbines provides an alternative strategy. Small components and personnel could be transferred from onshore directly onto turbines removing the need for an offshore-based operations centre. Larger components would need to be transported to site by boat, but this could be accomplished over several days without presenting any significant logistical challenge. Opportunities also exist for improving the availability of strategic spares stored on the turbine or offshore substation, in order to further optimise operations over the life of the farm.

The survey revealed that the use of offshore O&M platforms or the deployment of large O&M service vessels that remain on site for long durations, were thought to be the most likely industry standards that will emerge as offshore wind matures. Safety concerns over helicopter access in bad weather were identified as a major factor against this method of access.

O&M strategy and prioritisation

At the time of the survey the majority of large offshore wind farms were still within their OEM warranty period. During this period, maintenance of the assets is primarily the responsibility of the OEM. The operators have limited involvement, which is often limited to the specification of preferred condition monitoring systems and having technicians present to witness and support maintenance activities by the OEM.

O&M responsibility for the asset transfers away from the OEM at the end of the warranty period. There are a number of different models for this. Opinions vary and the survey did not reveal any one particular best practice approach currently emerging within the industry. The most popular approach for onshore and offshore wind farms to date has involved some level of shared O&M responsibility between the operator and the OEM in the post-warranty period. The balance of responsibilities varies with the size of the farm and the experience of the operator. Operators who already have significant experience of onshore O&M are often comfortable with transferring this knowledge to the offshore environment, and often take principal ownership of offshore O&M activity post-warranty period.

Maintenance strategies for wind farms must be highly tailored to meet the individual requirements for each farm. The OEM or the operator may choose to implement different methods in order to establish an optimal approach for minimising the impact of maintenance interventions on the cost of energy for each farm. The logistics of deploying maintenance teams far offshore and the access restrictions imposed by wind and the sea-state, dictates that more strategic maintenance practices will be required for far-offshore sites compared with the methodologies used for current wind farm operations. This will necessitate a further move away from reactive single-task maintenance activities based on condition monitoring, SCADA systems and regular inspection methodologies that were initially developed for onshore farms, towards preventative multi-element fixed-period component overhaul and replacement regimes. Maintenance bases for near-shore farms tend to be small operations carrying minimal inventory of replacement components and with low staff numbers. The survey revealed a clear consensus concerning the need to develop much larger offshore maintenance bases in order to meet the needs of the UK Round 3 sites. This includes the need to support strengthening of the supply chain to reduce cost and delivery time for OEM and third-party replacement components.

As UK offshore experience has grown, planned maintenance strategies have started to become more prevalent. Maintenance strategies for far-offshore farms have yet to be determined, but it was clear from the survey that more advanced condition monitoring, SCADA systems and preventative maintenance methodologies will be needed to meet cost of energy targets. Prioritising maintenance interventions will also become increasingly important as the industry develops. Far-offshore Round 3 wind farms will typically comprise of many hundreds of turbines and so the prognosis of emerging faults will become increasingly important in operational models for prioritising and scheduling of maintenance activity. Part of this will involve developments to improve operator confidence in condition monitoring and SCADA alarms.

O&M systems for far-offshore farms will need to enable early and accurate detection, diagnosis and prognosis of faults as they emerge. The risk of secondary damage following a component failure must also be assessed. The results from the remaining life calculations and risk assessment will be used to schedule and prioritise the maintenance action required, which will not always need to be carried out at the first opportunity. Reactive maintenance will only be undertaken if a failure is predicted to occur before the next scheduled maintenance opportunity. Temporary de-rating of turbines to prolong life or turbine shutdown to mitigate the risk of failure will become more commonplace in the optimisation of large wind farm systems for minimum the cost of energy.

Several of the stakeholders surveyed expressed the opinion that many of the solutions to the O&M challenges ahead lie in intelligent application and incremental development of existing techniques. For example, applying data mining techniques to condition monitoring and SCADA data in order to make comparisons between the operation of different turbines and farms will allow indices to be developed for efficiently monitoring the health of large numbers of wind turbine assets.

Health monitoring

SCADA and condition monitoring systems play a vital role in successfully operating remote wind farms. The health monitoring challenges identified in the survey included improving the reliability of monitoring systems and better communication of data. The greater distances to shore for future farm developments presents particular challenges for restricting spurious data or breaks in connection to acceptable levels. The cost of communication will be significant due to the numbers of turbines involved and the ever-increasing number of measurements per turbine. There can be over 200 signals to transmit from a turbine. In order to minimise costs it is important that turbines are not over-instrumented. Determining what to monitor, particularly in the early and late stages of the asset's life cycle, is critical to the development of systems, as excessive data can lead to communication and storage issues as well as information overload, resulting in operators routinely ignoring some data streams.

The development of intelligent alarm and data handling systems, will result in suitable responses being generated when faults emerge. These systems will be informed by past events and maintenance logs as experience grows. The aim will be for the system to facilitate early identification of what the alarm means, whether the turbine can continue to operate at full load or in a de-rated condition, and the risk of delaying maintenance intervention until the next scheduled maintenance opportunity.

Several commercial SCADA and condition monitoring systems exist that are moving towards these goals, but these are not yet standardised. At present, the lack of standardisation results in the need to develop expensive bespoke procedures and potential confusion for the operator. An operator with assets from several OEMs is likely to encounter multiple monitoring systems, each with their own complexities and limitations. The need for more rigorous standardisation of asset health monitoring systems emerged from the survey as a major opportunity for reducing the cost of energy from wind farms.

Conclusions

A literature review and a survey of stakeholders has been carried out to investigate the changing owner/operator requirements and the future O&M challenges for UK wind power as the industry grows and moves towards larger far-offshore wind farms. The cost of energy from existing UK offshore wind is too high and improving O&M operations will make an important contribution to reducing costs to sustainable levels as well as improving the safety of workers. The next generation UK Round 3 sites will see the development of wind farms consisting of hundreds of turbines often located at distances up to and beyond 100 km from shore. This will increase O&M challenges in relation to reduced maintenance opportunities, access difficulties, increased communication distances, larger SCADA and condition monitoring data volumes, increased lead times, more complex transport and storage of spare parts etc. The development of advanced new scheduled maintenance strategies and turbine health monitoring techniques is essential, in order to reduce costs and the reliance on expensive reactive maintenance practices. Much development work is currently being undertaken to move towards these goals. The need for better standardisation of systems and parts has also been identified from the survey to be an important area for future costs reduction.

Simulation and evaluation of wind farm operation

Introduction

An operation and maintenance (O&M) simulation tool has been developed and programmed in MATLAB© under the SUPERGEN Wind programme, for evaluating and estimating the Levelised Cost of Energy (LCoE) through simulating wind turbine failures and maintenance strategies over the lifetime of the wind farm. The O&M tool essentially consists of two main components, namely a wind farm simulator and a cost evaluation tool. The wind farm simulator models turbine random failures depending on the turbine topology and schedules maintenance and repairs for failed turbines depending on the type of faults, their severity and the type of maintenance strategy chosen. The wind farm simulator is essentially tasked with evaluating the levelised energy output of the wind farm over the course of its lifetime. Apart from the wind farm simulator the cost estimator is then used to estimate the LCoE of the wind farm based on the farm's levelised energy output. The cost estimator also takes into account the Capital Expenditure (CAPEX) as well as the variable Operational Expenditure (OPEX).

Monte Carlo Simulation (MCS) is used to sample the random failures of turbines based on reliability data down to the subassembly level for each turbine.

A fictitious typical wind farm arrangement, Exemplar Wind Farm (EWF), has been defined during the SUPERGEN Wind project and this has been used as the test case for evaluating the economics of wind farm operation in the work described in this section. The EWF is described in the next section. The performance of the EWF has been evaluated in terms of overall time-based availability of the farm, capacity factor (%), and LCoE (£/MWh) for four different O&M test case scenarios. These are described in the final part of this section along with the results and major conclusions from the study.

Wind farm simulator

As mentioned above the O&M simulation tool contains two main components. In this section the wind farm simulator is explained in more detail.

For the purposes of modelling it is assumed that all wind turbine generators are in their normal operating phase with constant time-independent state transition rates. The state probabilities can therefore be assumed to be exponential distributions and their random behaviour can be modelled as Markov processes. The variable state transition rate phases during 'infancy' and 'wear-out' are not included in the model developed during this study.

Figure 7.3 illustrates a two-state Markov process. Each subassembly in the turbine is modelled as a two-state Markov process on its own.

λ_{ij} and λ_{ji} are the state transition rates between the ith and the jth state and vice versa. These transition rates can either be failure or repair rates in the model.

Assuming an exponential distribution for the transition states, the turbine failures may be modelled as a multi-state time discrete Markov chain as shown in Figure 7.4. The states shown in Figure 7.4 are defined in Table 7.1.

In addition to these states, during maintenance the turbines with unhealthy components are shut-down and are temporarily placed in a maintenance state, in which further state sampling is not allowed until the affected turbine has undergone the necessary maintenance and repairs. Consequently during the

Figure 7.3 Two-state Markov chain model of a turbine sub-assembly

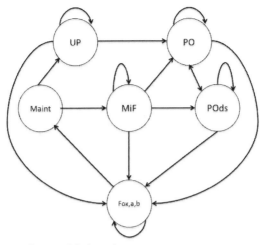

Figure 7.4 Turbine multi-state Markov chain

Table 7.1 Description of turbine states (El-Thalji (2012) and Alsyouf (2008))

State	Description
Up	Component is healthy
MiF	Maintenance induced failure state – similar to Up state but with slightly higher probability of failure
PO	Pre-outage state – a condition of having CMS installed, CMS has picked up an alarm, increased likelihood of failure
POds	Pre-outage de-rated state – rated power reduced to reduce the failure likelihood
FOa	Forced outage state – for type A failures
FOb	Forced outage state – for type B failures
FOx	Forced outage state – unknown

period of repairs the affected turbines would produce zero output power. There are only two possible outcomes (at random) from the maintenance state, namely a transition to Up state or to the MiF state. Once in MiF according to Table 7.1, the probability of a jump to an unhealthy state increases slightly.

It should be noted that in this study the specific type of the turbine is not considered, rather, three turbine failures types are considered, namely type A, B and X. Type A failures require small repairs and may or may not involve component replacements, however these may all be carried out with small support vessels and minimum crew numbers. All maintenance nodes are capable of carrying out repairs associated with type A failures. Type B failures, on the other hand, require the use of heavy vessels, and may also require provisions for replacement of the failed heavy component(s). Type X failure events correspond to unknown failures where an alarm has been raised by the Condition Monitoring System (CMS) however the nature of the failure is not identified. Type X failure events are verified only after a crew inspection to the affected turbine. After crew inspection has been carried out the crew then decides if the failure is a type A or type B. In the simulation this is done by a simple two-state Monte Carlo simulation in which there is a fixed probability for the outcome of the inspection.

Evaluation of wind farm availability and capacity factor

Each failure state of the wind turbines is sampled randomly through the Monte Carlo Simulation (MCS) over several thousand iterations. If the state of the wind turbine is called S_t the unbiased probability of that state after running N_{it} iterations can be determined for a wind farm with N_{turb} as:

$$P(S_t) = \frac{N(S_t)}{N_{turb} \times N_{it}} \tag{7.1}$$

where $N(S_t)$ is the number of occurrences of state S_t during the MCS process.

Similarly indices for wind farm availability can be defined as the probability of finding the wind farm at an available state at some point in time. The average power output of the wind farm is the total sum of the power outputs of all available wind turbines over the number of iterations, given as:

$$C_{WF}(i) = \sum_{n}^{N_{turb}} C_{turb}(i)_n \tag{7.2}$$

where $C_{turb}(i)_n$ is the power of the nth available turbine at the ith iteration which is a function of the sampled wind speed. When the farm's power output is calculated, turbine wake interaction effects have been ignored and it is assumed that all turbines rotate at the same speed which is determined by the wind speed at that point in time. The wind speed model thus does not allow for any wind speed variation across the farm (Byon 2011). It is also assumed that all turbines have maximum power point tracking capability that allows them to generate maximum power at any given wind speed (Mohan 2003).

Finally, the wind farm's Capacity Factor, CF_{WF} (%), is calculated as the average output of the wind farm over its theoretical output:

$$CF_{WF} = \frac{C_{WF}}{C_{WF}^{nom}} \times 100 \tag{7.3}$$

Economic Cost of Energy (CoE) model

The levelised CoE (LCoE) is the average cost of the energy generated by any kind of power plant over its operating lifetime (DECC 2013). For wind, since it is a capital intensive technology, the LCoE typically is orders of magnitude higher than conventional generation plant (DECC 2013). The cost model presented in this section has been adopted from the DECC–Mott Macdonald Cost Model

(DMCM) (DECC 2013, Mott Macdonald 2010). The DMCM contains the fixed initial Capital Expenditure (CAPEX) in £/kW for developing wind farms. Three groupings are considered, namely low, medium and high CAPEX, which depend upon factors such as turbine location and generating capacity (size). The cost model also includes the O&M costs associated with the running of the wind farm. O&M costs consist of fixed and variable costs. The fixed costs are estimated to be on average 2.7 per cent of the total fixed CAPEX (Mott Macdonald 2010). The DMCM either does not take into account variable O&M costs, or assumes an average value for the variable O&M costs over the operating lifetime of the turbine/farm. In the tool being described here, the focus is on the development of a mathematical time-dependent framework to model and simulate O&M activities on a day-by-day basis. So, the high fidelity dynamic modelling used in this tool allows the impact of, for example, different maintenance intervention strategies on the LCoE from the farm to be assessed – something which is not possible with averaging models such as the DMCM.

The LCoE is defined as the ratio of the present discounted value of the total incurred costs associated with owning and running a typical generating asset over the present discounted value of its total expected energy output, as shown in equation (7.4) below (DECC 2013).

$$LCoE = \frac{NPV_C}{NPV_G} \tag{7.4}$$

where NPV_C is the Net Present Value (NPV), or the present discounted value of the future incurred costs of owning and running the asset and NPV_G is the NPV of expected energy production from the asset (Mott Macdonald 2010). Typically the net energy generation is the gross energy production minus any internal or auxiliary use for the plant itself. In the current model no allowance has been made for auxiliary generation.

The LCoE can be thought of as the minimum average selling price of the energy generated by the asset to recover all of the operating expenses with zero profit.

A discount rate of 10 per cent, as recommended by DECC (Mott Macdonald 2010), has been adopted when calculating the NPVs. The NPVs are calculated in the model using:

$$NPV_C = \sum_i K_i^d (CAPEX + OPEX)_i \tag{7.5}$$

$$NPV_G = \sum_i K_i^d G_i \tag{7.6}$$

where K_i^d is the discount factor (for the ith year), $CAPEX_i$ is the asset's associated capital expenditure in year i and $OPEX_i$ is its operating expenditure during the year. CAPEX is principally incurred during the early years of planning and construction of farms. Once turbines have been commissioned OPEX becomes the major expenditure.

The SUPERGEN-Wind exemplar wind farm

The SUPERGEN-Wind Exemplar Wind Farm (EWF) is an imaginary example of a typical UK Round 3 offshore wind farm located on the Dogger Bank, UK. It was proposed and used by the SUPERGEN Wind Consortium as a standard test case that is free of any commercial constraints (SUPERGEN-Wind). The Dogger Bank development site is located 125–195 km off the East Yorkshire coast. It is the largest proposed development zone in UK Round 3 offshore wind expansion, with a potential capacity of up to 13 GW (Forewind). The EWF is populated by the SUPERGEN Exemplar Wind Turbine (EWT). This is a 5 MW turbine with a rotor diameter of 126 metres. The EWF specifications are given in Table 7.2. The 256 turbines are arranged in 16 rows each consisting of 16 turbines and covering an area of 195 km². In this work the EWF has been nominally assumed to be located 195 km offshore when simulating O&M activities.

The EWF package also includes an Exemplar Wind Speed (EWS) time series. The model wind speed distribution and directional wind rose were created using a mesoscale model and climatological data that is representative of typical UK Round 3 conditions.

Turbine failure simulation process

The turbine failure states are sampled within a Non-sequential state sampling Monte Carlo Simulation (NMCS) process according to transition probability shown in equation (7.7), assuming an exponential distribution for the state probabilities:

$$Q(t) = 1 - \exp(-\lambda t) \tag{7.7}$$

where λ is the state transition rate associated with the particular state transitions. More information about the underlying model and the NMCS process is given in Kazemtabrizi (2012) and Neate (2014) and will not be repeated here. It should be noted that the transition rates from MiF states are slightly higher than from the Up state to account for the higher risks of failure due to careless repair. As with all Markovian processes, it is necessary to account for the fact that every state transition depends only on its immediate preceding states and the system history

Table 7.2 SUPERGEN Wind exemplar wind farm specifications

EWF specification	
Total capacity	1.28 GW
Number of turbines	256
Turbine rating	5 MW
Turbine rotor diameter (D)	126 m
Turbine spacing within a row	6D
Row spacing	8D

is neglected. It goes without saying that in order to successfully implement the NMCS random numbers drawn from uniform distributions are generated for each component and for each state. The NMCS is essentially different from the Sequential Monte Carlo Simulation (SMCS) presented in Neate (2014) in which instead of sampling system states, state residence time durations were sampled.

Maintenance model

Once a component has failed it will be flagged by the maintenance model. Depending on the type of strategy chosen, and the weather model, a maintenance window is calculated for each day and the components are repaired only within that particular maintenance window. If a component cannot be repaired within a particular maintenance window, then the work will be carried forward to the next available maintenance window. Calculating the maintenance windows depends strongly on the type of the weather model and the maintenance strategy selected.

Proposed maintenance strategies

Four maintenance strategies based on developed strategies in Neate (2014) have been considered in this report, three of which are vessel-based strategies, which typically would involve hiring a large mobile vessel, such as a large boat, with the capacity to support smaller boats, and a helicopter. A fourth option, a platform-based maintenance strategy, has also been considered. In the fourth option, the maintenance crew are housed permanently on an offshore platform, which can either be a customised platform, or a remodelled offshore transformer platform. The main characteristics associated with each maintenance strategy are summarised in Table 7.3.

Options (1) and (2) utilise large boats, with a capacity to support smaller vessels for carrying out smaller repair tasks. Both options house permanent repair crews, with two-week rotations. Option (2) has a smaller capacity of operation in terms of small boats as well as an incapability for helicopter repairs and inspections due to lack of existence of a helipad, however, it has the highest speed amongst the vessel-based options.

The third option, which is an alternative to Options (1) and (2) would require jack-up boats, which typically are used for turbine installations, to remain behind and act as O&M support vessels. It has the lowest speed of the vessel-based options, and higher costs, however it has the highest operational capacity compared with other options, but no helicopter support. Option (4) would require adding an additional guard boat for heavy repair support, however in normal circumstances it is capable of supporting up to four smaller boats as well as a helicopter.

Costs are best assumptions and are based on data available in Neate (2014). The total O&M depends on the simulated activities which themselves depend on the type of maintenance strategy chosen. It is noted from data in Table 7.3 that net O&M costs of Option (2) would be the lowest of the options and Option (4) would be highest. This is in fact true, however when coupled with the energy

Table 7.3 Four maintenance strategies (cost data from Alsyouf, 2008)

Maintenance option	Vessel options			Platform option
	Option 1	Option 2	Option 3	Option 4
Node type	Large boat	Guard boat	Jack-Up	Offshore platform
Failure type B	Y	Y	Y	Y
Requires extra guard boat for failure type B?	N	N	N	Y
Failure B max wave height (m)	2.5	1.5	1	1.5
Speed (km/h)	20	25	10	0
Small boat carrying capacity	4	2	4	4
Crew size	2	2	4	1
Helicopter	Y	N	N	Y
Vessel charter (£/month)	10000	8500	12000	87500
Crew wages (£/month)	3500	3500	3500	3500

output and average availability of the farm which are derived from simulations, it is seen that the overall cost of energy (both undiscounted and levelised) associated with Option (2) is actually higher than for Option (4). This is because on average, Option (2) would yield a lower availability and capacity factor than Option (4) due to a smaller operational capacity.

Simulation results (cost of energy)

In this section the results of simulations are explained briefly. A base-case scenario has been considered in which the EWF is simulated with characteristics illustrated in Table 7.2. The base scenario places the EWF 195 km offshore covering a 195 km² surface area. The four maintenance strategies are simulated for a typical year and repeated 20 times to allow for an acceptable standard deviation in the final simulation results and to capture the unlikely random events. Accordingly, each scenario is simulated for an overall simulation time of 20 years. The total simulation time will then be 120 simulation years.

Essentially, once all system simulations are carried out, the total calculated OPEX is fed into the CoE model, prepared in Excel, which along with the fixed CAPEX calculates the discounted (levelised) cost of energy for each maintenance scenario. Apart from the OPEX costs calculated by the O&M model, as mentioned earlier, it also provides measures for the overall availability of the EWF for each maintenance strategy as well as an EWF Capacity Factor (CF). The CF is used to calculate the overall energy production of the farm over its useful life time period which spans for 22 years (from initial deployment in 2013 to final decommissioning in 2035), also taking into account the turbine degradation over time.

Base-case scenario

A base-case scenario for EWF with nominal values given in Table 7.2 has been run on MATLAB to calculate the average cost of O&M activities in the farm. The results are summarised in Figures 7.5 to 7.7.

It is noted from Figure 7.7 that Option (2) yields the lowest O&M costs, as percentage of total discounted costs (i.e. OPEX + CAPEX), however, it should be noted that the total discounted O&M costs, as calculated by simulating O&M activities, is not a reliable measure for determining the best maintenance strategy. The reason for lower O&M costs in Option (2) is simply due to the characteristics associated with that particular option, namely a smaller number of boats, which

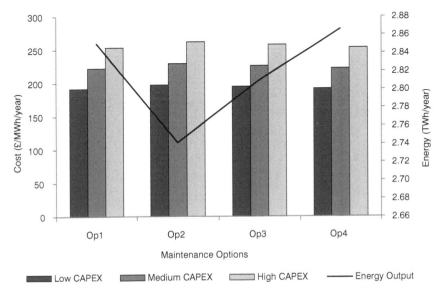

Figure 7.5 LCoE calculation for SUPERGEN-Wind EWF (base case)

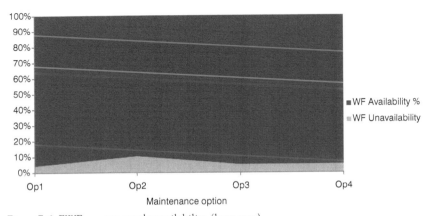

Figure 7.6 EWF average yearly availability (base case)

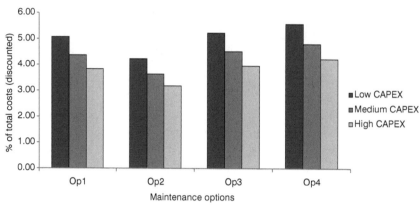

Figure 7.7 Total discounted O&M costs (per cent of total costs)

essentially leads to a smaller capacity of operation, naturally yielding a lower cost of operation and maintenance. However, as can be seen from Figure 7.5, this option is in effect the most costly option, since it also yields a lower average availability and energy output.

None of the options have managed to yield a cost as low as an average LCoE associated with combined cycle gas turbine (CCGT), which is taken to be £94/MWh in this study (Parsons 2010). It can be easily seen here that comparing results given in Figures 7.5 to 7.7, Options (1) and (4) are the best available options.

Even though Option (4) has a high cost of running, it nevertheless yields higher energy output, which eventually leads to a lower LCoE. The results suggest a strong negative correlation between the calculated LCoE and the average capacity factor (CF per cent) of the farm, which would further support the fact that higher availability would essentially be advantageous for most decreasing the overall LCoE. Consequently, the net costs of O&M are not a suitable measure for optimising strategy aimed at lowering the cost of energy associated with offshore wind.

Both options produce an overall LCoE between 191 and 253 £/MWh for a low to high CAPEX scenario and for a first-of-a-kind wind farm design and deployment scenario (Parsons 2010). This is compared with results given in Parsons (2010), which put the overall LCoE of a typical Round 3 offshore wind farm (EWF can be thought of as a typical Round 3 offshore wind farm) at somewhere between 180–200 £/MWh. The LCoE associated with each option is recalculated for CAPEX data for an nth-of-a-kind (mature) scenario to be between 160 and 230 £/MWh. Again Option (2) would be the most expensive option with an overall LCoE of 230 £/MWh compared with Options (1), (3) and (4), assuming the same levels of energy output. The decrease in overall LCoE is shown in Figure 7.8. It is solely the result of a decrease in CAPEX.

Optimising maintenance strategy – operational capacity

In order to optimise the maintenance strategy for the farm, sensitivity studies have been carried out by varying the key parameters associated with each maintenance

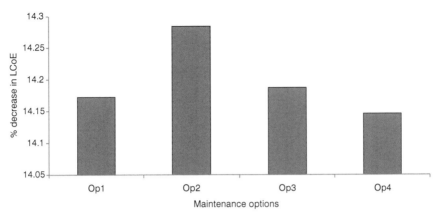

Figure 7.8 Percentage decrease in LCoE for nth-of-a-kind scenario (medium CAPEX – EWF)

option, namely, operational capacity, speed, and helicopter capability. Preliminary studies carried out using the O&M simulation tool have shown that there is a strong correlation between the net costs of O&M and the wind farm layout and distance to shore (i.e. larger, farther offshore wind farms will incur higher net O&M costs) (Neate 2014). However, the purpose of this study is to optimise the maintenance scenario subject to its key characteristics with the aim of minimising the LCoE. Option (2), as an example, has been chosen for further sensitivity studies. Option (4), the platform-based option, has already been shown to be comparable to Option (1), the best performing option, and therefore will not be studied here any further.

To further improve the performance of Option (2), its operational capacity has been doubled to support up to four small boats. Simulations were run again for the 20-year duration in order to calculate the O&M costs. The calculated LCoE values associated with Option (2) for both default and improved operational capacity are shown in Figure 7.9.

Moreover, results show that just by doubling the operational capacity of Option (2), the availability of the wind farm will increase dramatically compared with the default scenario. This would in effect increase the CF per cent of the farm which will essentially decrease the overall discounted cost of energy (LCoE) associated with Option (2) by 20 per cent. Again, it can be seen that by aiming for maximising energy output or availability of the wind farm, the overall LCoE will decrease dramatically.

The O&M code along with the CoE model may be combined together with a suitable optimisation toolbox for a more dynamic simulation approach. This would be a potential area for further research into optimising and quantifying O&M effects in larger offshore wind farms.

Conclusion

The performance of an offshore wind farm from the perspective of reliability and economics of operation has been considered in this chapter. In this work,

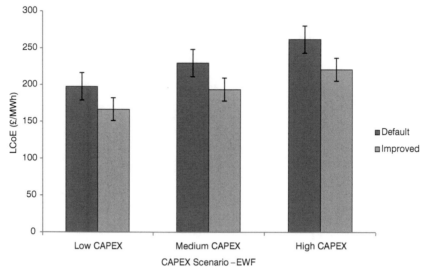

Figure 7.9 Effects of improving operational capacity of Option (2) on LCoE of EWF

the economics of performance of a typical offshore wind farm (SUPERGEN-Wind Exemplar) has been investigated. Results obtained from cost of energy calculations suggest a strong relationship between average availability and CF per cent of the wind farm and its overall levelised cost of energy (LCoE). This essentially suggests that for an efficient optimum maintenance strategy, the operator should aim for maximum availability in the farm in order to maximise its energy output. The sensitivity of the LCoE to different key characteristics of the offshore wind farm is shown clearly by these results.

References

Alsyouf, I., and El-Thalji, I., 2008, "Maintenance Practices in Wind Power Systems: a Review and Analysis", European Wind Energy Conference & Exhibition, Brussels.

Amirat, Y., Benbouzid, M.E.H., Al-Ahmar, E., Bensaker, B., and Turri, S., 2009, "A Brief Status on Condition Monitoring and Fault Diagnosis in Wind Energy Conversion Systems", *Renewable and Sustainable Energy Reviews*, 3(9), pp. 2629–2636.

Andrawus, J.A., Watson, J., Kishk, M., and Adam A., 2006, "A Selection of a Suitable maintenance Strategy for Wind Turbines", *Wind Engineering*, 30(6), pp. 471–486.

Besnard, F., Patrikssony, P., Strombergy, S., Wojciechowskiy, A., and Bertling, L., 2009, "An Optimization Framework for Opportunistic Maintenance of Offshore Wind Power System", Power Tech, Bucharest.

Besnard, F., Fischer, K., and Bertling, L., 2010a, "Reliability-Centred Asset Maintenance – A step Towards Enhanced Reliability, Availability, and Profitability of Wind Power Plants", Innovative Smart Grid Technologies Conference Europe, Gothenburg.

Besnard, F., and Bertling, L., 2010b, "An Approach for Condition-Based Maintenance Optimization Applied to Wind Turbine Blades", *IEEE Transactions on Sustainable Energy*, 1(2), pp. 77–83.

Byon, E., Ntaimo, L., and Ding, Y., 2010a, "Optimal Maintenance Strategies for Wind Turbine Systems Under Stochastic Weather Conditions", *IEEE Transactions on Reliability*, 59(2), pp. 393–404.

Byon, E., and Ding, Y., 2010b, "Season-Dependent Condition-Based Maintenance for a Wind Turbine Using a Partially Observed Markov Decision Process", *IEEE Transactions on Power Systems*, 25(4), pp. 1823–1834.

Byon, E., Perez, E., Ding., Y., and Ntaimo, L., 2011, "Simulation of Wind Farm Operations and Maintenance using DEVS", *International Transaction of the Society for Modeling and Simulation*, 87(12), pp. 1093–1117.

Crabtree, C.J., "Operational and Reliability Analysis of Offshore Wind Farms", EWEA 2012, Copenhagen.

Department of Energy and Climate Change (DECC), 2013, "Electricity Generation Costs", available at: www.gov.uk

Eecen, P.J., Braam, H., Rademakers, L.W.M.M., and Obdam, T.S., 2007, "Estimating Costs of Operations and Maintenance of Offshore Wind Farms", European Wind Energy Conference & Exhibition, Milan.

El-Thalji, I., and Liyanage, J.P., 2010, "Integrated Asset Management Practices for Offshore Wind Power Industry: A Critical Review and a Road Map to the Future", Proceedings of the Twentieth International Offshore and Polar Engineering Conference, Beijing.

El-Thalji, I., and Liyanage, J.P., 2012, "On the Operation and Maintenance Practices of Wind Power Asset: A Status Review and Observations", *Journal of Quality in Maintenance Engineering*, 18(3), pp. 232–266.

Forewind Consortium, Dogger Bank Development Zone, available at: http://www.forewind.co.uk/dogger-bank/overview.html

Hameed, Z., Hong, Y.S., Cho, Y.M., Ahn, S.H., and Song, C.K., 2009, "Condition Monitoring and Fault Detection of Wind Turbines and Related Algorithms: A Review", *Renewable and Sustainable Energy Reviews*, 13(1), pp. 1–39.

Hameed, Z., Vatn, J., and Heggset, J., 2011, "Challenges in the Reliability and Maintainability Data Collection for Offshore Wind Turbines", *Renewable Energy*, 36(8), pp. 2154–2165.

Harman, K., Walker, R., and Wilkinson, M., 2008, "Availability Trends Observed at Operational Wind Farms", European Wind Energy Conference & Exhibition, Brussels.

Kazemtabrizi, B., 2012, Monte Carlo Simulation for Large Offshore Wind Farms, SUPERGEN-wind deliverable for task 4.1.1. available from www.supergen-wind.org.uk

Lange, M., Wilkinson, M., van Delft. T., 2010, "Wind Turbine Reliability Analysis", Proceedings German Wind Energy Conference, Bremen.

Lu, B., Li, Y., Wu, X., and Yang, Z., 2009, "A Review of Recent Advances in Wind Turbine Condition Monitoring and Fault Diagnosis", Power Electronics and Machines in Wind Applications Conference, Lincoln, NE.

Marseguerra, M., Zio, Z., and Podofillinin, L., 2002, "Condition-Based Maintenance Optimization by Means of Genetic Algorithms and Monte Carlo Simulation", *Reliability Engineering and System Safety*, 77, pp. 151–166.

McMillan, D., and Ault, G.W., 2007, "Towards Quantification of Condition Monitoring Benefit for Wind Turbine Generators", European Wind Energy Conference & Exhibition, Milan.

McMillan, D., and Ault, G.W., 2008, "Condition Monitoring Benefit for Onshore Wind Turbines: Sensitivity to Operational Parameters", *IET Renewable Power Generation*, 2(1), pp. 60–72.

Mohan, N., Undeland, T.M. and Robbins, P.W., 2003, *Power Electronics: Converters, Applications, and Design*, Wiley, Chap. 17.

Mott Macdonald, 2010, "UK Electricity Generation Costs Update", Technical Report, available at: www.gov.uk

Neate, R., Kazemtabrizi, B., Golysheva, E., Crabtree, C., and Matthews, P., 2014, "Optimisation of Far Offshore Wind Farm Operation and Maintenance (O&M) Strategies", Full paper accepted as poster presentation at EWEA 2014.

Nielson, J.J., and Sorensen, J.D., 2010, "On Risk-Based Operation and Maintenance Planning for Offshore Wind Turbine Components", *Reliability Engineering & System Safety*, 96(1), pp. 218–229.

Nilsson, J., and Bertling, L., 2007, "Maintenance Management of Wind Power Systems Using Condition Monitoring Systems—Life Cycle Cost Analysis for Two Case Studies", *IEEE Transactions on Energy Conversion*, 22(1), pp. 223–229.

Parsons Brinckerhoff, 2010, "Powering the Nation", Technical Report. Parsons Brinckerhoff, London.

Rademakers, L.W.M.M., Braam, H., Zaaijer, M.B., and van Bussel, G.J.W., 2003, "Assessment and Optimisation of Operational and maintenance of Offshore Wind Turbines", European Wind Energy Conference & Exhibition, Madrid.

Ribrant, J., and Bertling, L., 2007, "Survey of Failures in Wind Power Systems With Focus on Swedish Wind Power Plants During 1997–2005", Power Engineering Society General Meeting, Tampa, FL.

Sørensen, J.D., 2007, "Optimal, Risk-Based Operation and Maintenance Planning for Offshore Wind Turbines", Proceedings of the European Offshore Wind Conference, Berlin.

Spinato, F., Tavner, P.J., van Bussel, G.J.W., and Koutoulakos, E., 2008, "Reliability of Wind Turbine Subassemblies", *IET Renewable Power Generation*, 3(4), pp. 387–401.

SUPERGEN-Wind Exemplar Wind Farm, available to consortium members at: www.SUPERGEN-wind.org.uk

Tavner, P.J., Xiang, J., and Spinato, F., 2007, "Reliability Analysis for Wind Turbines", *Wind Energy*, 10, pp. 1–18.

Utne, I.B., 2010, "Maintenance Strategies For Deep-Sea Offshore Wind Turbines", *Journal of Quality in Maintenance Engineering*, 16(4), pp. 367–381.

van Bussel, G.J.W., and Bierbooms, W.A.A.M., 2003, "Analysis of Different Means of Transport in the Operation and Maintenance Strategy for the Reference DOWEC Offshore Wind Farm", Proceedings of the European Seminar on Offshore Wind Energy in Mediterranean and Other European Seas, Naples, Italy.

van de Pieterman, R.P., Braam, H., Obdam, T.S., Rademakers, L.W.M.M., and van der Zee, T.J.J., 2011, "Optimisation of maintenance strategies for offshore wind farms. A case study performed with the OMCE-Calculator", Offshore 2011, Amsterdam.

Watson, S.J., and Xiang, J., 2006, "Real-Time Condition Monitoring of Offshore Wind Turbines", European Wind Energy Conference & Exhibition, Greece.

Wilkinson, M., Spinato, F., and Tavner, P.J., 2007, "Condition Monitoring of Generators & Other Subassemblies in Wind Turbine Drive Trains", 6th IEEE International Symposium on Diagnostics for Electric Machines, Power Electronics and Drives, Cracow.

Wilkinson, M., Hendriks, B., Spinato, F., Gomez, E., Bulacio, H., Roca, J., Tavner, P., Feng, Y., and Long, H., 2010, "Methodology and Results of the Reliawind Reliability Field Study", European Wind Energy Conference, Warsaw.

Wilkinson, M., 2011, "Measuring Wind Turbine Reliability – Results of the Reliawind Project", EWEA, Brussels.

Index

Note: When the text is within a table, the number span is in *italic* and when the text is within a figure, the number span is in **bold**.